Nonlinear Physical Science

Nonlinear Physical Science focuses on recent advances of fundamental theories and principles, analytical and symbolic approaches, as well as computational techniques in nonlinear physical science and nonlinear mathematics with engineering applications.

Topics of interest in *Nonlinear Physical Science* include but are not limited to:

- New findings and discoveries in nonlinear physics and mathematics
- Nonlinearity, complexity and mathematical structures in nonlinear physics
- Nonlinear phenomena and observations in nature and engineering
- Computational methods and theories in complex systems
- Lie group analysis, new theories and principles in mathematical modeling
- Stability, bifurcation, chaos and fractals in physical science and engineering
- Discontinuity, synchronization and natural complexity in physical sciences
- Nonlinear chemical and biological physics

This book series is indexed by the SCOPUS database.

To submit a proposal or request further information, please contact Dr. Mengchu Huang (Email: mengchu.huang@springer.com).

Jian-Qiao Sun

Random Vibration with Machine Learning Method

Higher Education Press

Springer

Jian-Qiao Sun
Department of Mechanical Engineering
University of California
Merced, CA, USA

ISSN 1867-8440 ISSN 1867-8459 (electronic)
Nonlinear Physical Science
ISBN 978-981-95-0811-2 ISBN 978-981-95-0812-9 (eBook)
https://doi.org/10.1007/978-981-95-0812-9

Jointly published with Higher Education Press Limited Company
The print edition is not for sale in the mainland of China. Customers from the mainland of China please order the print book from: Higher Education Press Limited Company.

This Springer imprint is published by the registered company Springer Nature Singapore Pte Ltd.
The registered company address is: 152 Beach Road, #21-01/04 Gateway East, Singapore 189721, Singapore

I dedicate this book to my Ph.D. advisor, the late Prof. C. S. Hsu (May 27, 1922–July 25, 2014), who introduced me to random vibration. His dedication, humility, and wisdom inspired me throughout my academic career.

Preface

Since my first book *Stochastic Dynamics and Control* was published in 2006 [1], the world has seen rapid advances in machine learning and artificial intelligence. Machine learning has penetrated many scientific and engineering disciplines, and is changing the way the engineering research is done. This book presents an updated volume on random vibration and includes some advances of the field in connection to machine learning.

The book briefly reviews fundamental theory of and methods for random vibration that can also be found in the existing books on the subject [1–9].

More important, the book reports the recent progress of applications of data-driven and machine learning methods to classical random vibration problems of nonlinear dynamic systems.

This book starts with very brief discussions on the theory of probability, stochastic process, stochastic calculus, response process as transformation and distribution of random responses. The method of neural networks is then introduced for the solution of random vibration problems including transient and steady state response probability density function, response moments and first passage time probability. Classical methods for solutions to random vibration problems of linear and nonlinear systems are briefly discussed. As the method of machine learning becomes more powerful and readily available, highly accurate numerical solutions of various random vibration problems can be obtained in the form of neural networks. Several chapters are dedicated to the topics on machine learning, Green's operator, Faynman-Kac formula, data driven random vibration analysis and new application of Monte Carlo simulation.

This book can serve as a reference on advanced studies of random vibration for active researchers. It also intends to serve the targeted audience of this book, who are graduate students in engineering. Examples in the book are beneficial to their learning.

One unique feature of this book is that we raise lots of motivating questions when we present the theory of random vibration. We believe that in the age of artificial intelligence when the powerful machine learning tools become widely available, random vibration is still a field full of opportunities for further research and engineering applications.

I would like to acknowledge Dr. Xi Wang of Xi'an Jiaotong University for his contributions. The examples of random vibration problems with machine learning are drawn from several publications by him and his advisors, my colleagues, Prof. Ling Hong and Prof. Jun Jiang. Special thanks also goes to my collaborate Prof. Zigang Li of Xi'an University of Science and Technology, whose work on data-driven generalized cell mapping is cited in the book and is a significant contribution to the field, and to my collaborator Prof. Lincong Chen of Huaqiao University, whose students, Dr. Zi Yuan and Dr. Jiamin Qian, contributed the materials in two chapters.

Merced, USA Jian-Qiao Sun
June 2025

References

1. Sun JQ (2006) Stochastic dynamics and control. Elsevier Science Ltd., Oxford, U.K
2. Cai G, Zhu WQ (2017) Elements of stochastic dynamics. World Scientific, Hackensack, New Jersey
3. Elishakoff I (1983) Probabilistic methods in the theory of structures. Wiley, New York
4. Ibrahim RA (1985) Parametric random vibration. Wiley, New York
5. Lin YK (1967) Probabilistic theory of structural dynamics. McGraw-Hill, New York
6. Lin YK, Cai GQ (1995) Probabilistic structural dynamics - advanced theory and applications. McGraw-Hill, New York
7. Soong TT (1973) Random differential equations in science and engineering. Academic Press, New York
8. Soong TT, Grigoriu M (1993) Random vibration of mechanical and structural systems. Prentice Hall, Englewood Cliffs, New Jersey
9. Wirsching PH, Paez TL, Ortiz H (1995) Random vibrations - theory and practice.Wiley, New York

Contents

Chapter 1
Introduction

1.1 When Are Uncertainties Random?

Uncertainties can be modeled as random variables or stochastic processes when there is a sufficient amount of data to form a sample space. Random uncertainties may arise from external excitations to the system, or from system parameters. Structural and mechanical systems can subject to earthquakes, ocean waves, wind loading and aerodynamic forces of fast moving vehicles, and from randomness of material parameters such as inaccurate modulus, internal material damping and geometrical variations due to variabilities of manufacturing processes. The subject of random vibration deals with response analysis for dynamical systems with random uncertainties and under random excitations.

It is commonly recognized that Einstein [1] established the theoretical foundation of stochastic calculus in his doctoral research and a series of papers on this subject, although Brownian motions, as a fundamental building block of stochastic calculus, were studied long before him [2]. The literature on stochastic dynamics has been steadily increasing in various subject areas including economy, finance, geology, ecology, population dynamics and biology. In the area of mechanical and civil systems, stochastic dynamics is known as random vibrations. The pioneering research work and the resulting book by Crandall and Mark [3] marked the beginning of the theory of random vibrations in engineering community. Since then, there have been quite a few excellent books on random vibrations [4–14].

1.2 Why Do We Study Random Vibration?

An important goal of structural analysis is to investigate the reliability of the structure subject to random loadings, and predict the probability of structural fatigue failure over a certain service time. Prediction of random vibration in time domain

J.-Q. Sun, *Random Vibration with Machine Learning Method*,
Nonlinear Physical Science, https://doi.org/10.1007/978-981-95-0812-9_1

or frequency domain combined with a fatigue model such as SN curve or fracture mechanics provides a basis for reliability design of structures against fatigue failure [15–22].

Mechanical systems are often designed to operate in a bounded domain with bounded velocities. Rolling motion of ship in random sea is an example. When the motion is out of the design domain for the first time, the system fails. The study of this problem is known as the first-passage time and reliability against first-passage failure [23–25].

The mathematical theory and solution methodologies of random vibrations are important part of knowledge base for mechanical, civil and naval engineers in their professional of design, analysis and optimization while taking into account of random uncertainties to develop a probabilistic assessment of reliability of mechanical systems [26–30].

The spectral analysis of random response and excitation is a classical topic. See Sect. 2.3. The correlation function describes cause-effect relationships of signals in measured random data, and has been used for diagnosis of mechanical systems [31]. For linear systems, the correlation function can be obtained in analytical form. In general, the correlation function is usually computed from data.

1.3 What Is Random Vibration About?

The theory of random vibration consists of the following topics, which are covered in [32].

- Time domain response prediction by means of Monte Carlo simulations
- Equations for response moments of linear systems and closure methods for nonlinear system
- Response spectral density by means of Fourier transform to study frequency content of random response
- Correlation analysis to investigate cause-effect relationship
- Response probability distribution by means of Fokker-Planck-Kolmogorov equation
- First-passage and reliability by means of backward Kolmogorov equation.

These topics are built on the foundation of probability theory and calculus of stochastic process and are discussed in [32]. This book presents a brief summary of these topics only. The theory helps us to obtain response statistics when the system model is known in terms of governing ordinary or partial differential equations. Various analytical solutions of response and probability distribution can be obtained from the equations for linear dynamic systems and for a limited class of nonlinear dynamic systems. See Chap. 7.

When we don't have equations for the system and we only have data, or when the system is nonlinear and complex so that no analytical solution can be found, the data driven approach becomes highly appealing. Some response statistics such as response moments, power spectrum and estimate of response probability distribution can be directly obtained from the data. Advanced machine learning methods using neural networks can also be adopted to solve for nonlinear partial differential equation to obtain transient and steady-state probability density function of the response for highly complex and high dimensional nonlinear systems. See Chaps. 8–12.

1.4 New Possibility with Random Vibration

Machine learning and artificial intelligence make use of large sets of data to build models for prediction and classification. For complex engineering systems such as manufacturing processes where motion and material property change happen at the same time, there is not enough real data currently. It is simply expensive to do comprehensive measurements of the system response and material changes in real time. Very often, physical changes of the system can be very difficult to directly measure even in the ideal laboratory environment. An example is the wear of wheel and rail of high-speed train.

Monte Carlo simulation of engineering systems with random uncertainties and excitations can provide large sets of data that are very useful for neural network modeling of the system. Random data are usually broadband with rich dynamic information about the system response, and therefore are better than deterministic response. In order for the simulated random data to be useful, high fidelity mathematical model of the engineering system must be developed and implemented digitally.

We should note that the simulated data and real-time measurements compliment each other. Combination of simulated data and real-time measurements in neural network modeling of complex engineering systems is in line with the current trend of AI research, i.e. data + knowledge, where the knowledge in terms of mathematical modal is the foundation of the simulated data.

It is therefore natural to anticipate that in the popular concept of digital twin, the model for a complex engineering system will have to consider random uncertainties and excitations. Hence, digital twin can potentially be an important part of random vibration in the future.

References

1. Einstein A (1956) Investigations on the theory of the Brownian movement. In: Furth R (ed). Dover Publications Inc., New York
2. Hänggi P, Marchesoni F, Nori F (2005) Brownian motors. Ann Phys 14(1–3):51–70
3. Crandall SH, Mark WD (1963) Random vibration in mechanical systems. Academic Press, New York
4. Lin YK (1967) Probabilistic theory of structural dynamics. McGraw-Hill, New York

5. Nigam NC (1983) Introduction to random vibrations. The MIT Press, Cambridge, Massachusetts
6. Newland DE (1984) Random vibrations and spectral analysis. Longman, New York
7. Ibrahim RA (1985) Parametric random vibration. Wiley, New York
8. Soong TT (1973) Random differential equations in science and engineering. Academic Press, New York
9. Young GE, Chang RJ (1988) Optimal control of stochastic parametrically and externally excited nonlinear control systems. J Dyn Syst Meas Contr 110:114–119
10. Soong TT, Grigoriu M (1993) Random vibration of mechanical and structural systems. Prentice Hall, Englewood Cliffs, New Jersey
11. Lin YK, Cai GQ (1995) Probabilistic structural dynamics - advanced theory and applications. McGraw-Hill, New York
12. Wirsching PH, Paez TL, Ortiz H (1995) Random vibrations - theory and practice. Wiley, New York
13. Elishakoff I (1983) Probabilistic methods in the theory of structures. Wiley, New York
14. Cai G, Zhu WQ (2017) Elements of stochastic dynamics. World Scientific, Hackensack, New Jersey
15. Wang X, Sun JQ (2005) On the fatigue analysis of non-gaussian stress processes with asymmetric distribution. J Vib Acoust 127:556–565
16. Wang X, Sun JQ (2005) Effect of skewness on fatigue life with mean stress correction. J Sound Vib 282:1231–1237
17. Wang X, Sun JQ (2005) Multi-stage regression fatigue analysis of non-gaussian stress processes. J Sound Vib 280(1):455–465
18. Wang X, Sun JQ (2003) Random fatigue of a higher order sandwich beam with system uncertainties. J Sound Vib 260(2):349–356
19. Wang X, Sun JQ (2002) Structural fatigue life prediction with system uncertainties. J Sound Vib 257(5):977–984
20. Sun JQ, Wang X, Bergman LA (2001) Fatigue analysis of nonlinear structures with von Mises stress. J Sound Vib 245(5):947–952
21. Sun JQ, Bao W, Miles RN (1998) Fatigue life prediction of nonlinear plates under random excitations. J Vib Acoust 120(2):353–360
22. Sun JQ, Miles RN (1991) Acoustic fatigue life prediction for non-linear structures. J Sound Vib 150(3):531–535
23. Sun JQ, Hsu CS (1988) First-passage time probability of non-linear stochastic systems by generalized cell mapping method. J Sound Vib 124(2):233–248
24. Wang X, Jiang J, Hong L, Sun JQ (2022a) First-passage problem in random vibrations with radial basis function neural networks. J Vib Acoust 144(5):051,014. https://doi.org/10.1115/1.4054437
25. Qian J, Chen L, Sun JQ (2024) A neural networks solution of first-passage problems. Appl Math Mech 45(11):2023–2036. https://doi.org/10.1007/s10483-024-3189-8
26. Denli H, Sun JQ, Chou TW (2005) Minimization of acoustic radiation from composite sandwich beam structures. AIAA J 43(11):2337–2341
27. Denli H, Sun JQ (2007) Structural-acoustic optimization of sandwich structures with cellular cores for minimum sound radiation. J Sound Vib 301(1–2):93–105
28. Denli H, Sun JQ (2007) Structural-acoustic optimization of composite sandwich structures: a review. Shock Vib Dig 39(3):189–200
29. Denli H, Sun JQ (2008) Structural acoustic optimization of sandwich cylindrical shells for minimum interior sound transmission. J Sound Vib 316(1–5):32–49
30. He MX, Sun JQ (2018) Multi-objective structural-acoustic optimization of beams made of functionally graded materials. Compos Struct 185:221–228. https://doi.org/10.1115/1.4036680
31. Bendat JS, Piersol AG (2000) Random data: analysis and measurement procedures. Wiley, New York
32. Sun JQ (2006) Stochastic dynamics and control. Elsevier Science Ltd., Oxford, U.K

Chapter 2
Theoretical Foundation

This chapter provides a brief review of several topics that form the theoretical foundation of random vibration. The topics include probability, stochastic process, stochastic calculus. For more in depth discussions of these topics, the readers are referred to the works discussed in the introduction chapter as well as the book [1].

2.1 Probability

The theory of probability and random variables is a classic topic. In this section, we merely discuss some important concepts that are used in the later chapters. For complete coverage of the topic, refer to the classical books on the topic [2, 3].

2.1.1 Random Variable

Uncertainties with many possible outcomes can be modeled as random variables. All possible outcomes form a *sample space* denoted by Ω.

Consider the displacement X of a structure subject to random excitations. X is a function of the random excitation. Formally, we define a *scalar random variable* $X(\omega)$ as a real function of the random sample $\omega \in \Omega$.

Let A_x denote a random event of the scalar random variable defined as

$$A_x = \{\omega \in \Omega : X < x\}. \tag{2.1}$$

A_x represents all random displacements less than x. A probability of the event is assigned as $P(A_x) = P(\omega \in \Omega : X < x)$.

J.-Q. Sun, *Random Vibration with Machine Learning Method*,
Nonlinear Physical Science, https://doi.org/10.1007/978-981-95-0812-9_2

It is common to drop the symbol ω in the argument of the random variable X for the sake of brevity. It is noted that random variables can form vectors and matrices.

2.1.2 Probability Distribution

Let $F_X(x)$ be the *distribution function* of X defined as

$$F_X(x) = P\left(A_x\right) = P(X \le x). \tag{2.2}$$

The distribution function is a *non-decreasing* function of x. In general, $F_X(x)$ is *continuous and piecewise differentiable*. The derivative of the distribution function is known as the *probability density function* (PDF) defined as

$$p_X(x) = \frac{d}{dx} F_X(x). \tag{2.3}$$

The probability density function is *nonnegative*, and satisfies the normalization condition in accordance with the probability axiom,

$$\int_{-\infty}^{\infty} p_X(x)\, dx = 1. \tag{2.4}$$

A random variable is completely specified by its probability density function, meaning that any statistics, such as the mean, variance, standard deviation, characteristic function and so on, of the random variable can be obtained from the probability density function.

Three common distributions of continuous and discrete random variables among many other are listed below.

1. **Uniform** $X \in U(a, b)$.

$$p_X(x) = \begin{cases} \frac{1}{b-a} & \text{if } a \le x \le b \\ 0 & \text{if } x < a \text{ or } x > b \end{cases} \tag{2.5}$$

2. **Normal or Gaussian** $X \in N(\mu, \sigma^2)$.

$$p_X(x) = \frac{1}{\sqrt{2\pi}\sigma} e^{-\frac{(x-\mu)^2}{2\sigma^2}}. \tag{2.6}$$

$$\mu_X = \mu,\ \sigma_X^2 = \sigma^2. \tag{2.7}$$

It is noted that the moments of the first two orders completely specify the probability density function, and therefore any other statistics of normal random variables.

3. **Poisson** $X \in P(\lambda)$. The probability of the random event $X = x$ is given by

$$P_P(X = x) = \frac{\lambda^x}{x!} e^{-\lambda}. \tag{2.8}$$

$$\mu_X = \sigma_X^2 = \lambda. \tag{2.9}$$

The time series of Poisson random variables is often used to approximate white noise excitations.

2.1.3 Function of Random Variable

Assume that $y = g(x)$ is a *differentiable monotonously increasing or decreasing* function of a scalar x. This guarantees a unique inversion $X = g^{-1}(Y)$. Note that

$$\int_{-\infty}^{\infty} p_Y(y)dy = \int_{-\infty}^{\infty} p_X(x)dx = 1. \tag{2.10}$$

Consider now the transformation of variable in the integration,

$$\int_{-\infty}^{\infty} p_X(x)dx = \int_{-\infty}^{\infty} p_X\left(g^{-1}(y)\right) \frac{1}{|g'\left(g^{-1}(y)\right)|} dy. \tag{2.11}$$

We have

$$p_Y(y) = p_X\left(g^{-1}(y)\right) \frac{1}{|g'\left(g^{-1}(y)\right)|}. \tag{2.12}$$

Consider now a multi-dimensional nonlinear *bijective (one-to-one) mapping*of random vectors

$$\mathbf{Y} = \mathbf{g}(\mathbf{X}), \tag{2.13}$$

where $\mathbf{X}$ and $\mathbf{Y}$ are n-dimensional random variables. The probability density function of $\mathbf{Y}$ is evaluated as

$$p_{\mathbf{Y}}(\mathbf{y}) = \frac{p_{\mathbf{X}}(\mathbf{g}^{-1}(\mathbf{y}))}{|J|}, \tag{2.14}$$

where

$$J = \det\left[\frac{\partial \mathbf{g}\left(\mathbf{g}^{-1}(\mathbf{y})\right)}{\partial \mathbf{x}^T}\right]. \tag{2.15}$$

is the *Jacobian* of the mapping, and $\mathbf{x} = \mathbf{g}^{-1}(\mathbf{y})$ is the inverse.

2.2 Stochastic Process

A *stochastic process* can be a time-indexed or space-indexed random variables. Here, we consider time-indexed continuous random variable $X(t)$. A continuous stochastic process $X(t)$ can be viewed as an infinite dimensional random vector.

The statistics of the stochastic process is described by the *joint probability density* or *distribution functions* of the random variables in the process.

2.2.1 Expectation of Stochastic Process

The *mean value function* $\mu_X(t)$ of the stochastic process $X(t)$ is defined as

$$\mu_X(t) = \mathrm{E}[X(t)] = \int_{-\infty}^{\infty} x p_X(x,t)dx, \tag{2.16}$$

where $p_X(x,t)$ is the probability density function of $X(t)$ at time t.

The *autocovariance function* of the stochastic process $X(t)$ at two time instances t_1 and t_2 is defined as

$$\begin{aligned}\kappa_{XX}(t_1,t_2) &= \mathrm{E}\left[(X(t_1) - \mu_X(t_1))\,(X(t_2) - \mu_X(t_2))\right] \\ &= \int_{-\infty}^{\infty}\int_{-\infty}^{\infty} (x_1 - \mu_X(t_1))\,(x_2 - \mu_X(t_2))\, p_X(x_1,t_1;x_2,t_2)dx_1dx_2 \\ &= R_{XX}(t_1,t_2) - \mu_X(t_1)\mu_X(t_2).\end{aligned} \tag{2.17}$$

Due to the symmetry of the expectation, it follows that

$$R_{XX}(t_1,t_2) = R_{XX}(t_2,t_1),\ \ \kappa_{XX}(t_1,t_2) = \kappa_{XX}(t_2,t_1). \tag{2.18}$$

When the process has zero mean, we have $R_{XX}(t_1,t_2) = \kappa_{XX}(t_1,t_2)$.

The *variance function* $\sigma_X^2(t)$ of the stochastic process at a given time is given by

$$\sigma_X^2(t) = \mathrm{E}\left[(X(t) - \mu_X(t))^2\right]. \tag{2.19}$$

The *autocorrelation coefficient function* $\rho_{XX}(t_1, t_2)$ of the process is defined as

$$\rho_{XX}(t_1, t_2) = \frac{\kappa_{XX}(t_1, t_2)}{\sigma_X(t_1)\sigma_X(t_2)}. \tag{2.20}$$

By using Schwarz inequality, we can show that $|\rho_{XX}(t_1, t_2)| \leq 1$ for any t_1 and t_2.

The nth order *moment functions* of $X(t)$ are defined as $\mathrm{E}[X(t)^n]$, and the corresponding *central moment functions* are defined as $\mathrm{E}[(X(t) - \mu_X(t))^n]$.

2.2.2 Characteristic and Cumulant Function

The characteristic function and the cumulant function of the stochastic process $X(t)$ can be defined in the same way as in the case of a single random variable. They are listed below.

$$M_X(\theta, t) \equiv \mathrm{E}[e^{i\theta X}] = \int_{-\infty}^{\infty} e^{i\theta x} p_X(x, t)dx. \tag{2.21}$$

Expanding the exponential term $e^{i\theta x}$ in power series, we have

$$M_X(\theta, t) = 1 + \sum_{n=1}^{\infty} \frac{(i\theta)^n}{n!}\mathrm{E}[X(t)^n]. \tag{2.22}$$

The nth order *moments* and *cumulant functions* of the stochastic process are given by

$$\mathrm{E}[X(t)^n] = \frac{1}{i^n}\frac{\partial^n M_X(\theta, t)}{\partial\theta^n}\bigg|_{\theta=0}, \tag{2.23}$$

$$k_n[X(t)] = \frac{1}{i^n}\frac{\partial^n \ln M_X(\theta, t)}{\partial\theta^n}\bigg|_{\theta=0}. \tag{2.24}$$

A stochastic process can also consist of a vector of random variables. Many concepts such as the mean, covariance, cross-covariance and cross-correlation coefficient can be defined in a similar way.

When a process consists of Gaussian random variables, it is called the Gaussian process. This is a popular process in engineering applications.

Several other concepts of stochastic processes include harmonic process, weak or strictly stationary process, ergodic process, Poisson process, compound Poisson process, Markov process. The references [4–10] have further discussions on stochastic processes.

2.3 Spectral Analysis

In engineering applications, we often need to know frequency content of stochastic processes. This study is known as spectral analysis [11–13].

Consider a *weakly stationary stochastic process with zero mean* such that the autocovariance and cross-covariance functions depend on the time difference. The Fourier transformation of the autocorrelation function $R_{XX}(\tau)$ of a scalar stochastic process $X(t)$ is known as the *power spectral density function* of $X(t)$.

$$S_{XX}(\omega) = \frac{1}{2\pi} \int_{-\infty}^{\infty} R_{XX}(\tau) e^{-i\omega\tau} d\tau. \tag{2.25}$$

The power spectral density function can be defined for an N-dimensional stochastic vector process. The result is a matrix.

In the frequency domain, several concepts are widely used in engineering, including coherence function, spectral moments and bandwidth. The classification of stochastic processes according to their bandwidth leads to the concepts of narrowband process, broadband process, band-limited white noise, and white noise.

2.4 Stochastic Calculus

Since stochastic processes are used to describe dynamics of engineering systems governed by differential equations, we must understand how to differentiate and integrate stochastic processes [7, 9, 10, 14, 15].

In the mathematics of calculus, the convergence of a series in some measure is the basis for the concepts such as continuity, differentiation and integration of functions. In this book, we focus on the convergence in mean square defined as

$$\lim_{n\to\infty} \mathrm{E}\left[|X_n - X|^2\right] = 0, \text{ or } \underset{n\to\infty}{\text{l.i.m.}} X_n = X. \tag{2.26}$$

Differentiation of a stochastic process $X(t)$ is defined in terms of the following limit

$$\dot{X}(t) = \underset{\Delta t \to 0}{\text{l.i.m.}} \frac{X(t + \Delta t) - X(t)}{\Delta t}, \tag{2.27}$$

in the mean square sense.

It can be shown that when the correlation function of the process $\frac{X(t+\Delta t)-X(t)}{\Delta t}$ exists and is finite as $\Delta t \to 0$, $\dot{X}(t)$ exists and the process $X(t)$ is differentiable in the mean square sense.

As a stochastic process $\dot{X}(t)$, its statistics and spectral properties can be evaluated when the probability density of it is available.

The derivative of Gaussian process and stationary process has interesting properties. More detail can be found in the reference books.

Integration of a stochastic process $\{X(t), t \in [a, b]\}$ can be defined in the following Riemann integral to generate a stochastic process $\{Y(t), t \in [a, b]\}$

$$Y(t) = \int_a^t h(t, \tau)X(\tau)d\tau. \tag{2.28}$$

The integration can be considered as the limit of the following random sequence

$$Y_n(t) = \sum_{k=1}^{n} h(t, \tau_k)X(\tau_k)\Delta\tau_k, \tag{2.29}$$

in the mean square sense, where $a = \tau_1 < \tau_2 < \cdots < t_n = t$ and $\Delta\tau_k = \tau_k - \tau_{k-1}$.

Various statistics of the integral including the mean, autocovariance function, and spectral properties can be computed in terms of the weight function $h(t, \tau)$ and the stochastic process $X(t)$.

Needless to say, differentiation and integration can be readily defined for vector stochastic processes.

2.4.1 Itô Calculus

Itô calculus deals with dynamical systems driven by Brownian motions such that the response is a Markov process [9, 14, 15].

The *Brownian motion* $B(t)$ is a special stochastic process, also known as the *Weiner process*, with the following interesting properties. It is a Gaussian process with zero mean, and a covariance function

$$\begin{aligned} R_{BB}(t, s) &= \kappa_{BB}(t, s) \\ &= \mathrm{E}[B(t)B(s)] = \sigma_B^2 \min(t, s). \end{aligned} \tag{2.30}$$

The Brownian motion has a few interesting properties. One of them is called Lévy's oscillation property, which leads to the following useful result in the mean square sense.

$$dB(t) \cdot dB(t) = |B(t + dt) - B(t)|^2 = \sigma_B^2 dt. \tag{2.31}$$

Consider a *Riemann-Stieltjes integral* defined as

$$Y(t) = \int_a^t f[X(\tau), \tau] dB(\tau), \tag{2.32}$$

where $X(t)$ is a stochastic process and $B(t)$ is a Brownian motion.

We discretize the time interval as $a = \tau_1 < \tau_2 < \cdots < \tau_n = t$. This integral can be viewed as the limit of a random sequence of the summation in the mean square sense.

In terms of the sequence, the *Itô integral* is defined as

$$Y_n^{\mathrm{I}}(t) = \sum_{k=1}^{n} f\left[X\left(\tau_{k-1}\right), \tau_{k-1}\right] [B(\tau_k) - B(\tau_{k-1})]. \tag{2.33}$$

It is noted that $f[X(\tau), \tau]$ is evaluated at the lower end of the time interval, and $dB(\tau)$ is taken as the forward difference. Because the increment $B(\tau_k) - B(\tau_{k-1})$ occurs after the time instance τ_{k-1}, it is independent of $X(\tau_{k-1})$.

The *Stratonovich integral* is defined as

$$Y_n^{\mathrm{S}}(t) = \sum_{k=1}^{n} f\left[X\left(\tau_{k-1/2}\right), \tau_{k-1/2}\right] [B(\tau_k) - B(\tau_{k-1})], \tag{2.34}$$

where $\tau_{k-1/2} = (\tau_k + \tau_{k-1})/2$.

The integration rules of conventional calculus are all applicable to the Stratonovich integral. Here, the process $X\left(\tau_{k-1/2}\right)$ may be dependent on the increment $B(\tau_k) - B(\tau_{k-1})$.

Corresponding to the integrals, we can also define *stochastic differential equations* in the sense of Itô or Stratonovich.

$$dX(t) = m(X, t)dt + \sigma(X, t)dB(t), \tag{2.35}$$

and its *formal solution*

$$X(t) = X(0) + \int_0^t m(X, \tau)d\tau + \int_0^t \sigma(X, \tau)dB(\tau). \tag{2.36}$$

From now on, we assume that $B(t)$ is a unit Brownian motion with $\sigma_B = 1$. One can view $\sigma(X, t)$ as a scaling factor that absorbs the parameter σ_B.

The difference between the Itô and Stratonovich integrals based on the stochastic equation is given by

$$\text{(S)} \int_0^t \sigma(X, \tau) dB(\tau) - \text{(I)} \int_0^t \sigma(X, \tau) dB(\tau) \tag{2.37}$$
$$= \frac{1}{2} \int_0^t \frac{\partial \sigma(X(\tau), \tau)}{\partial X} \sigma(X, \tau) d\tau.$$

This result helps us to convert the Stratonovich stochastic differential equation

$$\text{(S)}\ dX(t) = m(X, t)dt + \sigma(X, t)dB(t), \tag{2.38}$$

to the Itô differential equation

$$\text{(I)}\ dX(t) = \left[m(X, t) + \frac{1}{2} \frac{\partial \sigma(X, t)}{\partial X} \sigma(X, t) \right] dt \tag{2.39}$$
$$+ \sigma(X, t)dB(t),$$

where $\frac{1}{2} \frac{\partial \sigma(X,t)}{\partial X} \sigma(X, t)$ is called the *Wong-Zakai correction.*

For the nth dimensional vector stochastic process $X_j(t)$ satisfying the Stratonovich differential equation

$$\text{(S)}\ dX_j(t) = m_j(\mathbf{X}, t)dt + \sum_{k=1}^{m} \sigma_{jk}(\mathbf{X}, t)dB_k(t), \tag{2.40}$$

the corresponding Itô differential equation is given by

$$\text{(I)}\ dX_j(t) = m_j(\mathbf{X}, t)dt \tag{2.41}$$
$$+ \frac{1}{2} \sum_{k=1}^{n} \sum_{l=1}^{m} \frac{\partial \sigma_{jl}(\mathbf{X}, t)}{\partial X_k} \sigma_{kl}(\mathbf{X}, t)dt$$
$$+ \sum_{k=1}^{m} \sigma_{jk}(\mathbf{X}, t)dB_k(t),$$

where $dB_k(t)$ are increments of m independent unit Brownian motions.

When $\sigma_{jk}(\mathbf{X}, t)$ is not a function of $\mathbf{X}$, the Wong-Zakai correction vanishes, and the Itô and Stratonovich differential equations are completely equivalent. The most important result to carry forward is the Itô's lemma.

Assume that the nth dimensional vector stochastic process $X_j(t)$ satisfies the Itô differential equation

$$\text{(I) } dX_j(t) = m_j(\mathbf{X}, t)dt + \sum_{k=1}^{m} \sigma_{jk}(\mathbf{X}, t)dB_k(t). \tag{2.42}$$

Let $F(\mathbf{X}, t)$ be a function of $\mathbf{X}$ and t. Then, the differentiation of the function is given by the Taylor expansion of $F(\mathbf{X}, t)$ up to the order dt,

$$\begin{aligned} & dF(\mathbf{X}, t) \\ &= \left(\frac{\partial F}{\partial t} + \sum_{j=1}^{n} m_j \frac{\partial F}{\partial X_j} + \frac{1}{2} \sum_{j=1}^{n} \sum_{k=1}^{n} b_{jk} \frac{\partial^2 F}{\partial X_j \partial X_k} \right) dt \\ &+ \sum_{j=1}^{n} \sum_{k=1}^{m} \sigma_{jk} \frac{\partial F}{\partial X_j} dB_k(t), \end{aligned} \tag{2.43}$$

where

$$b_{jk} = \sum_{l=1}^{m} \sigma_{jl}\sigma_{kl}. \tag{2.44}$$

2.4.2 *Questions on Gaussian Property*

- Where is the Gaussian property required in the theory of Itô calculus?
- If the Brownian motion is non-Gaussian, would the theory of Itô calculus still hold?

References

1. Sun JQ (2006) Stochastic dynamics and control. Elsevier Science Ltd, Oxford, U.K
2. Feller W (1968) An introduction to probability theory and its applications, vol I. Wiley, New York
3. Feller W (1972) An introduction to probability theory and its applications, vol II. Wiley, New York
4. Doob JL (1953) Stochastic processes. Wiley, New York
5. Rice SO (1954) Mathematical analysis of random noise. In: Wax N (ed) Selected papers on noise and stochastic processes. Dover, New York, pp 133–294
6. Lin YK (1967) Probabilistic theory of structural dynamics. McGraw-Hill, New York
7. Wong E (1971) Stochastic processes in information and dynamical systems. McGraw-Hill, New York
8. Karlin S, Taylor HM (1981) A second course in stochastic processes. Academic Press, New York

9. Jazwinski AH (1970) Stochastic processes and filtering theory. Academic Press, New York
10. Sólnes J (1997) Stochastic processes and random vibrations - theory and practice. Wiley, New York
11. Shinozuka M, Deodatis G (1991) Simulation of stochastic processes by spectral representation. Appl Mech Rev 44(4):191–204
12. Newland DE (1984) Random vibrations and spectral analysis. Longman, New York
13. Bendat JS, Piersol AG (2000) Random data: analysis and measurement procedures. Wiley, New York
14. Lin YK, Cai GQ (1995) Probabilistic structural dynamics - advanced theory and applications. McGraw-Hill, New York
15. Soong TT (1973) Random differential equations in science and engineering. Academic Press, New York

Chapter 3
Simulation

3.1 Purpose of Simulation

Monte Carlo simulation starts with a mathematical model of the dynamic system and assumes that random uncertainties and excitations are specified with known probability distributions. Although a large set of random response data can be generated, Monte Carlo simulation as a research method is usually model-based.

The simulated random responses can be used for various purposes. The large sets of random responses give us statistics of the response including moments, correlation function, power spectral density, extreme events and probability distribution. For complex nonlinear dynamic systems with many degrees of freedom, Monte Carlo simulation has been the only effective method to obtain comprehensive response statistics, although it is computationally intensive. It is common to use Monte Carlo simulation to check the accuracy and validity of approximate analytical or numerical solutions for random vibration problem. Recently, applications of Monte Carlo simulation also appear to assist implementation of other analytical or numerical methods in a more effective manner. This will be studied in Chap. 13.

This chapter reviews the basic knowledge about how to simulate random numbers and stochastic processes.

3.2 Random Number

The most basic component of simulations is the generation of *pseudo-random numbers*. Uniformly distributed random numbers are usually first generated by the *linear congruential method* . This method is well documented in the textbook [1].

J.-Q. Sun, *Random Vibration with Machine Learning Method*,
Nonlinear Physical Science, https://doi.org/10.1007/978-981-95-0812-9_3

It should be pointed out that the formula to generate the uniform random numbers changes with the number of bits of the computer hardware.[1]

The uniform random numbers can be transformed to random variables with desired distribution including Gaussian, Rayleigh and so on. Independent uniform random numbers can be used to form vectors. Proper correlation structure of random vectors can be imposed by introducing linear transformation of the vector with independent components.

3.3 Random Process

When a random vector is indexed with respect to time or spatial coordinates, we simulate random process or random field.

Take the Gaussian white noise as an example. Note that the correlation of the white noise process in time domain is a delta function. Its power spectral density function is constant over the infinite frequency domain $[0, \infty)$. This is a mathematical idealization and cannot be realized with the simulated random numbers. In the simulation literature, there are various ways to approximate the Gaussian white noise. One realistic way for engineering applications is to introduce a bandwidth such that the power spectral density function is constant in the finite frequency range.

Other random processes are also often simulated including harmonic process, Poisson process and general color process. A very effective approach to simulate the random process with a pre-described power spectral density function is due to Shinozuka [2].

3.4 Integration of Stochastic Differential Equation

The response of dynamic systems subject to random excitations is usually governed by differential equations. To numerically simulate the response, we can convert the equations of motion in the state space form. The well-documented numerical integration algorithms such as Runge-Kutta algorithm can be used to integrate the equations subject to proper initial conditions.

Consider a nonlinear stochastic differential equation in the Stratonovich sense.

$$\begin{aligned} \dot{\mathbf{X}}(t) &= \mathbf{f}(\mathbf{X}, t) + \mathbf{W}(t), \\ \mathbf{X}(t) &= \mathbf{X}_0 \end{aligned} \tag{3.1}$$

where $\mathbf{X}$ is the state vector, $\mathbf{f}(\mathbf{X}, t)$ is a nonlinear function of its arguments, and $\mathbf{W}(t)$ is a vector of stochastic processes.

[1] https://en.wikipedia.org/wiki/Linear_congruential_generator.

Let $\mathbf{W}(t)$ be the simulated time series of the random excitation at a certain sample time interval. Let Δt be the numerical integration time step and $t_k = k\Delta t$. The fourth order Runge-Kutta algorithm is stated as

$$\begin{aligned}\mathbf{k}_1 &= \Delta t[\mathbf{f}(\mathbf{X}(k\Delta t), k\Delta t) + \mathbf{W}(k\Delta t)], \\ \mathbf{k}_2 &= \Delta t[\mathbf{f}(\mathbf{X}(k\Delta t) + \mathbf{k}_1/2, (k+1/2)\Delta t) + \mathbf{W}((k+1/2)\Delta t)], \\ \mathbf{k}_3 &= \Delta t[\mathbf{f}(\mathbf{X}(k\Delta t) + \mathbf{k}_2/2, (k+1/2)\Delta t) + \mathbf{W}((k+1/2)\Delta t)], \\ \mathbf{k}_4 &= \Delta t[\mathbf{f}(\mathbf{X}(k\Delta t) + \mathbf{k}_3, (k+1)\Delta t) + \mathbf{W}((k+1)\Delta t)],\end{aligned} \tag{3.2}$$

$$\mathbf{X}((k+1)\Delta t) = \mathbf{X}(k\Delta t) + \frac{1}{6}(\mathbf{k}_1 + 2\mathbf{k}_2 + 2\mathbf{k}_3 + \mathbf{k}_4) + O(\Delta t^5). \tag{3.3}$$

Note that in the fourth order method, if the sample time interval for the solution $\mathbf{X}(t)$ is Δt, then the sample time interval for the random process must be $\Delta t/2$ because middle points of $\mathbf{W}(t)$ at time instances $(k+1/2)\Delta t$ are used in the algorithm.

With the simulated response, we can calculate various statistics including the first-passage time probability and probability density function.

3.5 A Note on Non-Gaussian Process

Finally, we make a brief remark on simulation of non-Gaussian process. When a nonlinear transformation from a Gaussian variable to the specified non-Gaussian variable is given, it is quite straightforward to simulate the non-Gaussian variable with a pre-determined probability distribution.

When no such a transformation is available, the non-Gaussian probability distribution is often approximated with the help of higher order moments [3]. That is, we look for nonlinear transformations of Gaussian random variable to create non-Gaussian random variable such that the simulated non-Gaussian random variable would have a probability distribution that matches the specified probability distribution up to a certain order of moments.

In addition to the requirement for the probability distribution, a non-Gaussian process may also have a specified spectral distribution. This translates to the need for a temporal correlation structure that matches the spectral distribution. If we impose temporal correlation structure by means of linear transformations, we may cause deviation from the target probability distribution. As a result, simulation of non-Gaussian process is often iterative [4].

References

1. Knuth DE (1998) The art of computer programming. Addison-Wesley
2. Shinozuka M, Deodatis G (1991) Simulation of stochastic processes by spectral representation. Appl Mech Rev 44(4):191–204
3. Winterstein SR (1988) Nonlinear vibration models for extremes and fatigue. J Eng Mech 114(10):1772–1790
4. Grigoriu M (1998) Simulation of stationary non-Gaussian translation processes. J Eng Mech 124(2):121–126

Chapter 4
Response Process

In random vibration analysis, we are interested in the response of a dynamic system subject to random excitations or due to the influence of uncertain random parameters. In this chapter, we briefly discuss the response process as another example of stochastic processes. In particular, we adopt the view of the response as a transformation of random inputs. The concept and method of random variable transformation and stochastic processes can be readily seen as the statistical foundation of random vibration.

4.1 System with Random Uncertainty

Consider a general nonlinear dynamic system subject to random excitations and with parameter uncertainties. The uncertainties are modeled as random variables. For convenience, we assume that the system satisfies the following state equation.

$$\begin{aligned} \dot{\mathbf{X}}(t) &= \mathbf{F}(\mathbf{X}, \boldsymbol{\alpha}, \mathbf{W}(t)), \\ \mathbf{X}(0) &= \mathbf{X}_0, \end{aligned} \tag{4.1}$$

where $\mathbf{X}(t) \in \mathbb{R}^n$ is the response random vector process, $\boldsymbol{\alpha} \in \mathbb{R}^m$ is a constant random vector, representing all uncertain parameters, $\mathbf{W}(t) \in \mathbb{R}^l$ is a vector of random excitations and $\mathbf{X}_0$ is an initial condition, which can also be a random variable.

$$\mathbf{F}(\cdot) : \mathbb{R}^n \times \mathbb{R}^m \times \mathbb{R}^l \mapsto \mathbb{R}^n \tag{4.2}$$

is a nonlinear function of its arguments.

The response of the system can be formally written as

$$\mathbf{X}(t) = \mathcal{G}[\mathbf{X}_0, \boldsymbol{\alpha}, \mathbf{W}(t)], \tag{4.3}$$

J.-Q. Sun, *Random Vibration with Machine Learning Method*,
Nonlinear Physical Science, https://doi.org/10.1007/978-981-95-0812-9_4

where

$$\mathcal{G}: \mathbb{R}^n \times \mathbb{R}^m \times \mathbb{R}^l \mapsto \mathbb{R}^n. \tag{4.4}$$

The solution $\mathcal{G}$ is viewed as a transformation that maps the random vector $\{\mathbf{X}_0, \boldsymbol{\alpha}, \mathbf{W}(t)\}$ to the response process $\mathbf{X}(t)$.

Example 4.1 Consider a single degree of freedom system subject to random excitation and initial conditions.

$$\ddot{X} + 2\zeta\omega_n\dot{X} + \omega_n^2 X(t) = W(t), \tag{4.5}$$

$$X(0) = X_0, \quad \dot{X}(0) = V_0. \tag{4.6}$$

where the damping ratio $0 < \zeta < 1$. The response can be expressed in the following general form in mean square sense.

$$X(t) = X_h(t) + \int_0^t h(t-\tau)W(\tau)d\tau, \tag{4.7}$$

where $X_h(t)$ is the homogenous solution due to the initial conditions and $h(t)$ is the impulse response.

$$X_h(t) = e^{-\sigma t}\left(X_0 \cos\omega_d t + \frac{V_0 + \sigma X_0}{\omega_d}\sin\omega_d t\right), \tag{4.8}$$

$$h(t) = \begin{cases} \frac{1}{\omega_d} e^{-\sigma t}\sin\omega_d(t), & t \geq 0, \\ 0, & t < 0, \end{cases} \tag{4.9}$$

where $\sigma = \zeta\omega_n$ and $\omega_d = \omega_n\sqrt{1-\zeta^2}$.

Equation (4.7) clearly defines a relationship between the input random uncertainties $\{X_0, V_0, \zeta, \omega, W(t)\}$ to the output $X(t)$. It is also clear that the relationship between the random uncertain parameters $\{X_0, V_0, \zeta, \omega\}$ to $X(t)$ is algebraic and determined by the mathematical model of the system. In the language of machine learning, this is called the knowledge.

The relationship between $W(t)$ and $X(t)$ is a functional mapping determined by the impulse response function $h(t)$. Usually, $h(t)$ is also determined by the mathematical model of the system.

Consider a case when we don't have a model available. We do have the input output data $X(t_i)$ and $W(t_i)$ where $i = 0, 1, \ldots, N_s$. For linear systems with zero initial conditions, the relationship between the input and output is dictated by the convolution integral in Eq. (4.7), which can be discretized as

$$\hat{X}(t_i) = \sum_{k=0}^{i} h(t_i - t_k)W(t_k)\Delta t, \tag{4.10}$$

where $\hat{X}(t_i)$ is an estimate of $X(t_i)$, Δt is the sample time of the measurement and can be absorbed by $h(t)$. Let $h(t)$ be expressed in terms of neural networks with trainable coefficients $\mathbf{w}$.

Define a loss function in terms of the vector of prediction errors as

$$J(\mathbf{w}) = \frac{1}{2}\mathbf{e}^T\mathbf{e}, \tag{4.11}$$

$$\mathbf{e} = \{X(t_i) - \hat{X}(t_i)\} \in \mathbb{R}^{N_s \times 1}. \tag{4.12}$$

With a training method, we can find an optimal $\mathbf{w}$ to minimize the loss $J(\mathbf{w})$.

This last part of the example illustrates the concept of neural operators very popular in the literature [1]. The advantage of neural operator lies in that we only need to train a neural network model for the system once to obtain the impulse response function. The response of the system to an arbitrary excitation can be readily obtained by the convolutional operation in Eq. (4.10).

If the system parameters such as ζ and ω_n are unknown, can they be estimated in the data-driven approach? The answer is yes. We shall discuss this in Chap. 11.

4.2 Extended Sample Space

Let Ω_α denote the sample space for the random parameter $\boldsymbol{\alpha}$, $\Omega_\mathbf{W}$ the sample space of the random excitation $\mathbf{W}(t)$, $\Omega_\mathbf{X}$ the sample space of the random response $\mathbf{X}(t)$. Hence, we have

$$\Omega_\mathbf{X} = \Omega_\alpha \cap \Omega_\mathbf{W}. \tag{4.13}$$

We have the total probability of the response sample space

$$P(\Omega_\mathbf{X}) = P(\Omega_\alpha \cap \Omega_\mathbf{W}) = P(\Omega_\alpha)P(\Omega_\mathbf{W}) = 1. \tag{4.14}$$

In the literature, this is often referred to as the conservation of probability.

4.3 Nature of Response Process

As a stochastic process, random response of a dynamic system can be classified in the same manner. That is, the response can be stationary, weakly stationary, non-stationary, Markovian, non-Markovian and ergodic. The spectral properties and correlation function of the response can all be studied once the response process is obtained.

This is the subject of random vibration analysis and will be discussed in the following chapters.

4.4 Probability of Response

Assume that $\boldsymbol{\alpha}$ and $\mathbf{W}(t)$ are independent. Recall the transformations of random variables. Consider a case where there is no random excitation $\mathbf{W}(t) = 0$. Since $\mathbf{X}(t)$ and $\boldsymbol{\alpha}$ are related to each other by a nonlinear algebraic relationship, the probability density function of $\mathbf{X}(t)$ with respect to the uncertain parameters $\boldsymbol{\alpha}$ is determined by Eq. (2.14). Hence, when the solution is obtained and the transformation $\mathcal{G}$ is also specified, the probability density function of $\mathbf{X}(t)$ is fully determined.

When random excitations are active, we have to find another way to obtain response PDF. Let $p_{\mathbf{X}}(\mathbf{x}, t)$ be the probability density function of the response process, and $p_{\mathbf{X}\boldsymbol{\alpha}}(\mathbf{x}, \boldsymbol{\alpha}, t)$ the joint PDF of the response and the random parameter. Making use of the conditional probability formula, we have

$$p_{\mathbf{X}\boldsymbol{\alpha}}(\mathbf{x}, \boldsymbol{\alpha}, t) = p_{\mathbf{X}}(\mathbf{x}, t|\boldsymbol{\alpha}) p_{\boldsymbol{\alpha}}(\boldsymbol{\alpha}). \tag{4.15}$$

Hence,

$$\begin{aligned} p_{\mathbf{X}}(\mathbf{x}, t) &= \int_{\Omega_{\boldsymbol{\alpha}}} p_{\mathbf{X}\boldsymbol{\alpha}}(\mathbf{x}, \boldsymbol{\alpha}, t) d\boldsymbol{\alpha} \\ &= \int_{\Omega_{\boldsymbol{\alpha}}} p_{\mathbf{X}}(\mathbf{x}, t|\boldsymbol{\alpha}) p_{\boldsymbol{\alpha}}(\boldsymbol{\alpha}) d\boldsymbol{\alpha}. \end{aligned} \tag{4.16}$$

Note that $p_{\mathbf{X}}(\mathbf{x}, t|\boldsymbol{\alpha})$ is the response PDF to the time-dependent random excitations $\mathbf{W}(t)$ conditional on a given parameter $\boldsymbol{\alpha} \in \Omega_{\boldsymbol{\alpha}}$. How to obtain this conditional PDF is the subject of Chap. 6 on Fokker-Planck-Kolmogorov equation.

Reference

1. Li Z, Kovachki N, Azizzadenesheli K, Liu B, Bhattacharya K, Stuart A, Anandkumar A (2021) Fourier neural operator for parametric partial differential equations. arXiv:2010.08895

Chapter 5
Response Moment

5.1 Moment Equation

An important application of Itô's lemma is to derive a set of *moment equations* for the response of the Itô differential equation. Recall that in the Itô sense, $dB_k(t)$ is defined as the forward difference and $\sigma_{jk}(\mathbf{X}, t)$ is independent of $dB_k(t)$. Also, $\mathrm{E}[dB_k(t)] = 0$. Taking the mathematical expectation on both sides of Eq. (2.41), we have

$$\frac{d\mathrm{E}[X_j(t)]}{dt} = \mathrm{E}[m_j(\mathbf{X}, t)]. \tag{5.1}$$

Consider $F(\mathbf{X}, t) = X_j X_k$. According to Itô's lemma, we have

$$\begin{aligned} d(X_j X_k) &= \left(m_j X_k + m_k X_j + b_{jk}\right) dt \\ &\quad + \sum_{l=1}^{m} \left(\sigma_{jl} X_k + \sigma_{kl} X_j\right) dB_l(t). \end{aligned} \tag{5.2}$$

Taking the expectation of the equation, and recalling that

$$\mathrm{E}[\left(\sigma_{jl} X_k + \sigma_{kl} X_j\right) dB_l(t)] = 0, \tag{5.3}$$

because of the independence of the terms $\sigma_{jl} X_k + \sigma_{kl} X_j$ and $dB_l(t)$ in the Itô sense, we have the equation for the correlation function

$$\frac{d\mathrm{E}[X_j X_k]}{dt} = E\left[m_j X_k + m_k X_j + b_{jk}\right]. \tag{5.4}$$

By following the same steps, we can construct differential equations governing the evolution of the moments of any order.

J.-Q. Sun, *Random Vibration with Machine Learning Method*,
Nonlinear Physical Science, https://doi.org/10.1007/978-981-95-0812-9_5

Example 5.1 Consider a linear Itô differential equation

$$dX_j(t) = \sum_{k=1}^{n} A_{jk} X_k(t)dt + \sum_{k=1}^{m} \sigma_{jk} dB_k(t). \tag{5.5}$$

The equation governing the evolution of the mean is

$$\frac{d\mathrm{E}[X_j(t)]}{dt} = \sum_{k=1}^{n} A_{jk}\mathrm{E}[X_k(t)]. \tag{5.6}$$

In the vector form, it reads

$$\frac{d\mathrm{E}[\mathbf{X}]}{dt} = \mathbf{A}\mathrm{E}[\mathbf{X}]. \tag{5.7}$$

The correlation matrix satisfies the following equation

$$\frac{d\mathrm{E}[X_j X_k]}{dt} = \sum_{l=1}^{n} \left(A_{jl}\mathrm{E}[X_l X_k] + A_{kl}\mathrm{E}[X_l X_j]\right) + b_{jk}, \tag{5.8}$$

or

$$\frac{d\mathbf{R_{XX}}}{dt} = \mathbf{A}\mathbf{R_{XX}} + \mathbf{R_{XX}}\mathbf{A}^T + \mathbf{b}. \tag{5.9}$$

5.2 Cumulant Function

The method of Gaussian closure also leads to the relationship of higher order moments to lower order ones. Furthermore, in Gaussian closure, we are only interested in the first and second order response moments of the system, because the Gaussian distribution only needs the first and second order moments to be completely specified.

When the moments of order higher than two are to be obtained, we imply that the response will have a non-Gaussian distribution.

The characteristic function and cumulant function defined in Sect. 2.2.2 can be extended to a vector process $\mathbf{X}(t) \in \mathbb{R}^n$.

$$\begin{aligned} M_{\mathbf{X}}(\boldsymbol{\theta}, t) \equiv \mathrm{E}[e^{i\boldsymbol{\theta}^T\mathbf{X}}] &= \int_{\mathbb{R}^n} e^{i\boldsymbol{\theta}^T\mathbf{X}} p_{\mathbf{X}}(\mathbf{x}, t)d\mathbf{x} \\ &= 1 + \sum_{k=1}^{\infty} \frac{i^k}{k!} \left(\boldsymbol{\theta}^T\mathrm{E}[\mathbf{X}(t)]\right)^k, \end{aligned} \tag{5.10}$$

where $\boldsymbol{\theta} \in \mathbb{R}^n$.

The first to third order cumulant functions of the vector stochastic process are given by

$$k_1[\mathbf{X}] = \frac{1}{i}\frac{\partial \ln M_{\mathbf{X}}(\boldsymbol{\theta}, t)}{\partial \theta_j}\bigg|_{\boldsymbol{\theta}=\mathbf{0}}, \tag{5.11}$$

$$k_2[\mathbf{X}] = \frac{1}{i^2}\frac{\partial^2 \ln M_{\mathbf{X}}(\boldsymbol{\theta}, t)}{\partial \theta_j \theta_l}\bigg|_{\boldsymbol{\theta}=\mathbf{0}}, \tag{5.12}$$

$$k_3[\mathbf{X}] = \frac{1}{i^3}\frac{\partial^3 \ln M_{\mathbf{X}}(\boldsymbol{\theta}, t)}{\partial \theta_j \theta_l \theta_m}\bigg|_{\boldsymbol{\theta}=\mathbf{0}}, \tag{5.13}$$

where $1 \leq j, l, m \leq n$. Higher order cumulant functions can be constructed in the same manner. In the following, we list the moment relationships from third to fourth order closure.

Let us first introduce short-hand notations for the cumulant functions and moments.

$$k_3[X_i, X_j, X_k] = k_3[ijk], \quad \mathrm{E}[X_i^n X_j^m X_k^l] = \mathrm{E}[i^n j^m k^l]. \tag{5.14}$$

Third Order Cumulant

$$k_3[ijk] = \mathrm{E}[ijk] - \sum^{(3)} \mathrm{E}[i]\mathrm{E}[jk] + 2\mathrm{E}[i]\mathrm{E}[j]\mathrm{E}[k], \tag{5.15}$$

where $\sum^{(3)}$ indicates the sum of three different permutations of indices $[ijk]$. An example of resulting moment equations when $k_3[ijk] = 0$ reads

$$k_3[111] = \mathrm{E}[x_1^3] - 3\mathrm{E}[x_1]\mathrm{E}[x_1^2] + 2\mathrm{E}[x_1]^3 = 0. \tag{5.16}$$

That is $\mathrm{E}[x_1^3] = 3\mathrm{E}[x_1]\mathrm{E}[x_1^2] - 2\mathrm{E}[x_1]^3$.

Forth Order Cumulant

$$\begin{aligned} k_4[ijkl] = {} & \mathrm{E}[ijkl] - \sum^{(4)} \mathrm{E}[i]\mathrm{E}[jkl] + 2\sum^{(6)} \mathrm{E}[i]\mathrm{E}[j]\mathrm{E}[kl] \\ & - \sum^{(3)} \mathrm{E}[ij]\mathrm{E}[kl] - \sum^{(6)} \mathrm{E}[i]\mathrm{E}[j]\mathrm{E}[k]\mathrm{E}[l]. \end{aligned} \tag{5.17}$$

An example reads

$$k_4[1111] = \mathrm{E}[x_1^4] - 4\mathrm{E}[x_1]\mathrm{E}[x_1^3] + 12\mathrm{E}[x_1]^2\mathrm{E}[x_1^2] \\ - 3\mathrm{E}[x_1^2]^2 - 6\mathrm{E}[x_1]^4 = 0. \quad (5.18)$$

Hence, $\mathrm{E}[x_1^4] = 4\mathrm{E}[x_1]\mathrm{E}[x_1^3] - 12\mathrm{E}[x_1]^2\mathrm{E}[x_1^2] + 3\mathrm{E}[x_1^2]^2 + 6\mathrm{E}[x_1]^4$.

The readers can try to write down the explicit expressions for other third and fourth order cumulant functions.

5.3 Cumulant-Neglect Closure

When the terms $m_j(\mathbf{X}, t)$ and $\sigma_{jk}(\mathbf{X}, t)$ in Eq. (2.42) or in Eq. (5.4) are nonlinear functions of the state $\mathbf{X}$, moment equations are coupled and form an infinite hierarchy. We have to break the hierarchy in order to compute the time response of the moments.

There are two popular closure methods for studying the statistical properties of the responses of nonlinear systems. The first one is the method of non-Gaussian closure [1]. The idea of the method is to assume a non-Gaussian probability density function with adjustable parameters for the response and to use the moment relations derived from the system equations to obtain equations for the unknown parameters. The resulting probability density function can then provide approximate response statistics.

The second is the method of cumulant-neglect closure [2, 3]. The method simply neglects the cumulants of system variables above certain order in order to close the infinite hierarchy of the equations governing the moments of system variables.

With a nonlinear example, it has been shown in [4] that when the nonlinear system response is far from being Gaussian, higher order method of cumulant-neglect closure may lead to unreliable predictions of the response moment.

References

1. Crandall SH (1985) Non-gaussian closure techniques for stationary random vibration. Int J Non-Linear Mech 20(1):1–8
2. Wu WF, Lin YK (1984) Cumulant-neglect closure for non-linear oscillators under parametric and external excitations. Int J Non-Linear Mech 9(4):349–362
3. Ibrahim RA (1985) Parametric random vibration. Wiley, New York
4. Sun JQ, Hsu CS (1987) Cumulant-neglect closure methods for nonlinear systems under random excitations. J Appl Mech 54:649–655

Chapter 6
Response Distribution

The previous chapters have discussed relatively low order statistics such as mean, variance and correlation functions. The probability density function can determine all statistics of the response. Here, we consider the equation governing the evolution of probability density functions of the response processes of the stochastic differential equations [1–3]. The equation is known as Fokker-Planck-Kolmogorov (FPK) equation and can be derived in different ways. This chapter presents two approaches.

6.1 Governing Equation

The first and conventional approach to derive FPK equation is to make use of Chapman-Kolmogorov-Smoluchowski (CKS) equation. Probability density functions satisfy a so-called consistence equation

$$p_X(x_2, t_2|x_1, t_1) = \int_{R^1} p_X(x_2, t_2; x, t|x_1, t_1)dx. \tag{6.1}$$

Applying the formula for conditional probability

$$p_X(x_2, t_2; x, t|x_1, t_1) = p_X(x_2, t_2|x, t; x_1, t_1)p_X(x, t|x_1, t_1), \tag{6.2}$$

and assume that $t_2 > t > t_1$ and $X(t)$ is a stationary Markov process, we have

$$p_X(x_2, t_2|x_1, t_1) = \int_{R^1} p_X(x_2, t_2|x, t)p_X(x, t|x_1, t_1)dx. \tag{6.3}$$

This is the well-known *Chapman-Kolmogorov-Smoluchowski equation* The CKS equation can also be obtained for vector Markov processes.

J.-Q. Sun, *Random Vibration with Machine Learning Method*,
Nonlinear Physical Science, https://doi.org/10.1007/978-981-95-0812-9_6

Consider an infinitely differentiable and arbitrary function $R(y)$ such that

$$\lim_{y \to \pm\infty} \frac{d^n R(y)}{dy^n} = 0, \quad \text{for any } n > 0. \tag{6.4}$$

Consider an integration

$$\begin{aligned} I &= \int_{R^1} R(y) \frac{\partial}{\partial t} p_X(y, t|x_0, t_0) dy \\ &= \int_{R^1} R(y) \lim_{\Delta t \to 0} \frac{1}{\Delta t} \left(p_X(y, t + \Delta t|x_0, t_0) - p_X(y, t|x_0, t_0)\right) dy. \end{aligned} \tag{6.5}$$

where $t > t_0$.

Making use of the CKS equation and the fact that $R(x)$ is arbitrary, we can obtain

$$\frac{\partial p_X(x, t|x_0, t_0)}{\partial t} = -\frac{\partial}{\partial x}[A(x, t) p_X] + \frac{\partial^2}{\partial x^2}\left[\frac{B(x, t)}{2!} p_X\right] - \cdots \tag{6.6}$$

where

$$\begin{aligned} A(x, t) &= \lim_{\Delta t \to 0} \frac{1}{\Delta t} \mathrm{E}[\Delta X | X = x], \\ B(x, t) &= \lim_{\Delta t \to 0} \frac{1}{\Delta t} \mathrm{E}[\Delta X^2 | X = x]. \end{aligned} \tag{6.7}$$

A question can be raised about the truncation. When should we truncate the series expansion in Eq. (6.6)? This is partially answered by Pawula theorem [2].

Theorem 6.1 (Pawula Theorem) *In order for the probability density function $p_X(x, t|x_0, t_0)$ to be positive, the series may stop after the first or second term. If it does not stop after the second term, it must contain an infinite number of terms.*

We should point out that the Pawula theorem emphasizes the positivity of the probability density function only, and does not address the issue of convergence and solution accuracy of the truncated partial differential equation.

Example 6.1 The equation of motion of a linear oscillator can be written in a state space form

$$\begin{aligned} dX_1 &= X_2 dt, \\ dX_2 &= (-2\zeta\omega_0 X_2 - \omega_0^2 X_1) dt + d\beta(t), \end{aligned} \tag{6.8}$$

where $\beta(t)$ is a Wiener process with the following properties

$$\mathrm{E}[d\beta(t)] = 0, \ \mathrm{E}[d\beta(t) d\beta(s)] = \begin{cases} 2D dt, & t = s, \\ 0, & t \neq s. \end{cases} \tag{6.9}$$

The formal derivative of $\beta(t)$ is known as the white noise. Let $\mathbf{X} = [X_1, X_2]^T$ be the vector stochastic process of the system. According to Eq. (6.7), we have

$$A_1 = \lim_{\Delta t \to 0} \frac{1}{\Delta t} \mathrm{E}[\Delta X_1 | \mathbf{X} = \mathbf{x}] = x_2, \tag{6.10}$$

$$A_2 = \lim_{\Delta t \to 0} \frac{1}{\Delta t} \mathrm{E}[\Delta X_2 | \mathbf{X} = \mathbf{x}] = -2\zeta\omega_0 x_2 - \omega_0^2 x_1, \tag{6.11}$$

$$\begin{aligned} B_{11} &= \lim_{\Delta t \to 0} \frac{1}{\Delta t} \mathrm{E}[\Delta X_1^2 | \mathbf{X} = \mathbf{x}] \\ &= \lim_{\Delta t \to 0} \frac{\Delta t^2}{\Delta t} \mathrm{E}[X_2 | \mathbf{X} = \mathbf{x}] = 0, \end{aligned} \tag{6.12}$$

$$\begin{aligned} B_{21} = B_{12} &= \lim_{\Delta t \to 0} \frac{1}{\Delta t} \mathrm{E}[\Delta X_1 \Delta X_2 | \mathbf{X} = \mathbf{x}] \\ &= \lim_{\Delta t \to 0} \frac{1}{\Delta t} \mathrm{E}[\Delta t^2 X_2(-2\zeta\omega_0 X_2 - \omega_0^2 X_1) + \Delta t X_2 \Delta\beta(t) | \mathbf{X} = \mathbf{x}] \\ &= 0, \end{aligned} \tag{6.13}$$

$$\begin{aligned} B_{22} &= \lim_{\Delta t \to 0} \frac{1}{\Delta t} \mathrm{E}[\Delta X_2^2 | \mathbf{X} = \mathbf{x}] \\ &= \lim_{\Delta t \to 0} \frac{1}{\Delta t} \mathrm{E}[(-2\zeta\omega_0 X_2 - \omega_0^2 X_1)^2 \Delta t^2 \\ &\quad + 2(-2\zeta\omega_0 X_2 - \omega_0^2 X_1)\Delta t \Delta\beta(t) + \Delta\beta(t)^2 | \mathbf{X} = \mathbf{x}] = 2D. \end{aligned} \tag{6.14}$$

The FPK equation for the oscillator is given by

$$\begin{aligned} &\frac{\partial}{\partial t} p_{\mathbf{X}}(\mathbf{x}, t | \mathbf{x}_0, t_0) \\ &= -\frac{\partial}{\partial x_1}[A_1(\mathbf{x}, t) p_{\mathbf{X}}] - \frac{\partial}{\partial x_2}[A_2(\mathbf{x}, t) p_{\mathbf{X}}] + \frac{\partial^2}{\partial x_1^2}\left[\frac{B_{11}(x, t)}{2!} p_{\mathbf{X}}\right] \\ &\quad + 2\frac{\partial^2}{\partial x_1 \partial x_2}\left[\frac{B_{12}(x, t)}{2!} p_{\mathbf{X}}\right] + \frac{\partial^2}{\partial x_2^2}\left[\frac{B_{22}(x, t)}{2!} p_{\mathbf{X}}\right] \\ &= -\frac{\partial}{\partial x_1}[x_2 p_{\mathbf{X}}] - \frac{\partial}{\partial x_2}[(-2\zeta\omega_0 x_2 - \omega_0^2 x_1) p_{\mathbf{X}}] + D\frac{\partial^2 p_{\mathbf{X}}}{\partial x_2^2}. \end{aligned} \tag{6.15}$$

We now derive the FPK equation via Itô's lemma for an arbitrary function $F(\mathbf{X})$.

$$\begin{aligned} dF(\mathbf{X}) &= \left(\sum_{j=1}^{n} m_j \frac{\partial F}{\partial X_j} + \frac{1}{2} \sum_{j=1}^{n} \sum_{k=1}^{n} \sum_{l=1}^{m} \sigma_{jl}\sigma_{kl} \frac{\partial^2 F}{\partial X_j \partial X_k} \right) dt \\ &\quad + \sum_{j=1}^{n} \sum_{k=1}^{m} \sigma_{jk} \frac{\partial F}{\partial X_j} dB_k(t) + O(dt^{1.5}), \end{aligned} \tag{6.16}$$

where $dX_j(t) = m_j(\mathbf{X}, t)dt + \sum_{k=1}^{m} \sigma_{jk}(\mathbf{X}, t)dB_k(t)$ in the Itô sense and $\{B_k(t)\}$ is an m-dimensional vector of independent unit Brownian motions. Taking the mathematical expectation of Eq. (6.16) conditional on $(\mathbf{x}_0, t_0)$ and noting that

$$\sum_{j=1}^{n} \sum_{k=1}^{m} \mathrm{E}\left[\sigma_{jk} \frac{\partial F}{\partial X_j} dB_k(t) \middle| \mathbf{x}_0, t_0 \right] = 0, \tag{6.17}$$

due to Itô's interpretation, we have

$$\begin{aligned} &\frac{d}{dt} \int_{\mathbb{R}^n} F(\mathbf{x}) p_{\mathbf{X}}(\mathbf{x}, t|\mathbf{x}_0, t_0) d\mathbf{x} \\ &= \int_{\mathbb{R}^n} \left(\sum_{j=1}^{n} m_j \frac{\partial F}{\partial x_j} + \frac{1}{2} \sum_{j=1}^{n} \sum_{k=1}^{n} b_{jk} \frac{\partial^2 F}{\partial x_j \partial x_k} \right) p_{\mathbf{X}}(\mathbf{x}, t|\mathbf{x}_0, t_0) d\mathbf{x}, \end{aligned} \tag{6.18}$$

where $b_{jk} = \sum_{l=1}^{m} \sigma_{kl} \sigma_{jl}$.

Integrating by parts of the right hand side of Eq. (6.18) making an assumption that as $|x_k| \to \infty$, $p_{\mathbf{X}}(\mathbf{x}, t|\mathbf{x}_0, t_0)$ and its derivatives up to second order with respect to $\mathbf{x}$ all vanish, and making use of the fact that $F(\mathbf{x})$ is arbitrary, we obtain the FPK equation

$$\begin{aligned} \frac{\partial}{\partial t} p_{\mathbf{X}}(\mathbf{x}, t|\mathbf{x}_0, t_0) = &- \sum_{k=1}^{n} \frac{\partial}{\partial x_k} [m_k(\mathbf{x}, t) p_{\mathbf{X}}] \\ &+ \sum_{j=1}^{n} \sum_{k=1}^{n} \frac{\partial^2}{\partial x_j \partial x_k} \left[\frac{b_{jk}(\mathbf{x}, t)}{2!} p_{\mathbf{X}} \right]. \end{aligned} \tag{6.19}$$

This derivation indicates that the FPK is of order dt. The truncation of the series expansion in Eq. (6.6) at the second order leads to the same FPK equation and therefore has the same order dt of accuracy.

To fully determine the probability density function of the response, we must also specify the initial condition.

$$p_{\mathbf{X}}(\mathbf{x}, t_0|\mathbf{x}_0, t_0) = p_0(\mathbf{x}). \tag{6.20}$$

6.1.1 Questions on FPK Equation

There are several important and unanswered questions about the derivation of the FPK equation.

- In the two different ways to derive the FPK equation, it is assumed that in the limit when the state approaches infinity, the response PDF and its partial derivatives vanish. The assumption implies that the dynamic system is stable such that the probability for the response to reach infinity is zero.
- When the states of a system are constrained in a bounded region, how do we derive the FPK equation for this stochastic process?
- For the system constrained in the finite domain of the state space, it is common to take the FPK equation and impose necessary boundary conditions. Is this a correct way to handle the constrained system in a bounded region?
- Conversely, if we impose boundary conditions to the FPK equation, does it lead to the Itô equations with the correct state constraints?
- For colored noise excitation, the response is non-Markov. Additional state variables can be introduced to model the color noise as the output of an Itô equation. In the higher dimensional extended state space, the non-Markov response process becomes part of a Markov process. This is how we extend the theory to the non-Markov response of the system subject to colored noise excitation.
- In the derivation of the FPK equation, there is no discussion whether the excitation is Gaussian or non-Gaussian. Regardless, for nonlinear systems, the random vibration response is usually non-Gaussian. The entire derivation remains valid.

6.2 Backward Equation

The conditional probability density function of a Markov process $p_X(x, t|x_0, t_0)$ is a function of two sets of arguments, namely, (x, t) and (x_0, t_0). The FPK equation deals with the variation of the density function with respect to the arguments (x, t). This describes the forward dynamics of the system since $t > t_0$.

We now study how $p_X(x, t|x_0, t_0)$ changes with respect to (x_0, t_0). In other words, we are interested in the backward dynamics [1–3]. The backward dynamics allows us to estimate the reliability of the system.

Starting with the Chapman-Kolmogorov-Smoluchowski equation, we follow the same lines of derivation as for the FPK equation, we obtain the *backward Kolmogorov equation* governing the evolution of the conditional probability density function

$$-\frac{\partial}{\partial t_0} p_X(x, t|x_0, t_0) = A(x_0, t_0)\frac{\partial p_X}{\partial x_0} + \frac{B(x_0, t_0)}{2!}\frac{\partial^2 p_X}{\partial x_0^2} + \cdots \tag{6.21}$$

where

$$\begin{aligned} A(x_0, t_0) &= \lim_{\Delta t \to 0} \frac{1}{\Delta t}\mathrm{E}[\Delta X | X = x_0], \\ B(x_0, t_0) &= \lim_{\Delta t \to 0} \frac{1}{\Delta t}\mathrm{E}[\Delta X^2 | X = x_0]. \end{aligned} \tag{6.22}$$

Applying Theorem 6.1 (the Pawula theorem), we can keep the first two terms of the backward Kolmogorov equation leading to

$$-\frac{\partial}{\partial t_0}p_X(x,t|x_0,t_0)=A(x_0,t_0)\frac{\partial p_X}{\partial x_0}+\frac{B(x_0,t_0)}{2!}\frac{\partial^2 p_X}{\partial x_0^2}. \tag{6.23}$$

Example 6.2 The equation of motion of a linear oscillator excited by a white noise can be written in a state space form

$$\begin{aligned} dX_1 &= X_2 dt, \\ dX_2 &= (-2\zeta\omega_0 X_2-\omega_0^2 X_1)dt+d\beta(t), \end{aligned} \tag{6.24}$$

where $\beta(t)$ is a Wiener process with the following properties

$$\mathrm{E}[\beta(t)]=0,\ \mathrm{E}[\beta(t_1)\beta(t_2)]=2D(t_1-t_2),\quad t_1>t_2. \tag{6.25}$$

Let $\mathbf{X}=[X_1,X_2]^T$ be the vector stochastic process of the system. According to Eq. (6.22), we have

$$\begin{aligned} A_1 &= \lim_{\Delta t\to 0}\frac{1}{\Delta t}\mathrm{E}[\Delta X_1|\mathbf{X}=\mathbf{x}_0]=x_{02}, \\ A_2 &= \lim_{\Delta t\to 0}\frac{1}{\Delta t}\mathrm{E}[\Delta X_2|\mathbf{X}=\mathbf{x}_0]=-2\zeta\omega_0 x_{02}-\omega_0^2 x_{01}, \end{aligned} \tag{6.26}$$

$$\begin{aligned} B_{11} &= \lim_{\Delta t\to 0}\frac{1}{\Delta t}\mathrm{E}[\Delta X_1^2|\mathbf{X}=\mathbf{x}_0] \\ &= \lim_{\Delta t\to 0}\frac{\Delta t^2}{\Delta t}\mathrm{E}[X_2|\mathbf{X}=\mathbf{x}_0]=0, \\ B_{21} &= B_{12}=\lim_{\Delta t\to 0}\frac{1}{\Delta t}\mathrm{E}[\Delta X_1\Delta X_2|\mathbf{X}=\mathbf{x}_0] \\ &= \lim_{\Delta t\to 0}\frac{1}{\Delta t}\mathrm{E}[\Delta t^2 x_{02}(-2\zeta\omega_0 x_{02}-\omega_0^2 x_{01}) \\ &\quad +\Delta t x_{02}\Delta\beta(t)|\mathbf{X}=\mathbf{x}_0]=0, \\ B_{22} &= \lim_{\Delta t\to 0}\frac{1}{\Delta t}\mathrm{E}[\Delta X_2^2|\mathbf{X}=\mathbf{x}] \\ &= \lim_{\Delta t\to 0}\frac{1}{\Delta t}\mathrm{E}[(-2\zeta\omega_0 x_{02}-\omega_0^2 x_{01})^2\Delta t^2 \\ &\quad +2(-2\zeta\omega_0 x_{02}-\omega_0^2 x_{01})\Delta t\Delta\beta(t)+\Delta\beta(t)^2|\mathbf{X}=\mathbf{x}_0]=2D. \end{aligned} \tag{6.27}$$

The backward Kolmogorov equation for the oscillator is given by

$$-\frac{\partial}{\partial t_0}p_{\mathbf{X}}(\mathbf{x},t|\mathbf{x}_0,t_0)=x_{02}\frac{\partial p_{\mathbf{X}}}{\partial x_{01}} - (2\zeta\omega_0 x_{02}+\omega_0^2 x_{01})\frac{\partial p_{\mathbf{X}}}{\partial x_{02}}+D\frac{\partial^2 p_{\mathbf{X}}}{\partial x_{02}^2}. \tag{6.28}$$

6.3 Reliability Formulation

Let $\mathbf{X}(t)$ be an n-dimensional vector process. $\mathcal{S}\subseteq\mathbb{R}^n$ denotes a domain in which the system is considered to be safe. Γ is the boundary of $\mathcal{S}$. Assume that $\mathbf{X}(t_0)=\mathbf{x}_0\in\mathcal{S}$ at time t_0. The probability that the system is still in the *safe domain* $\mathcal{S}$ at time t is given by

$$\begin{aligned}\mathcal{R}_{\mathcal{S}}(t,t_0,\mathbf{x}_0)&=P\left(t<T\cap\mathbf{X}(t)\in\mathcal{S}|\mathbf{X}(t_0)=\mathbf{x}_0\right)\\&=\int_{\mathcal{S}}p_{\mathbf{X}}(\mathbf{x},t|\mathbf{x}_0,t_0)d\mathbf{x},\end{aligned} \tag{6.29}$$

where T is the first time when $\mathbf{X}(t)$ crosses the boundary Γ. $\mathcal{R}_{\mathcal{S}}(t,t_0,\mathbf{x}_0)$ is known as the reliability against the *first-passage failure* with respect to the safe domain $\mathcal{S}$.

Integrating the backward equation for a vector process over $\mathcal{S}$ with respect to $\mathbf{x}$, we obtain a partial differential equation of the reliability function $\mathcal{R}_{\mathcal{S}}(t,t_0,\mathbf{x}_0)$.

$$-\frac{\partial\mathcal{R}_{\mathcal{S}}(t,t_0,\mathbf{x}_0)}{\partial t_0}=\sum_{j=1}^{n}A_j(\mathbf{x}_0,t_0)\frac{\partial\mathcal{R}_{\mathcal{S}}(t,t_0,\mathbf{x}_0)}{\partial x_{0j}} +\sum_{j=1}^{n}\sum_{k=1}^{n}\frac{B_{jk}(\mathbf{x}_0,t_0)}{2!}\frac{\partial^2\mathcal{R}_{\mathcal{S}}(t,t_0,\mathbf{x}_0)}{\partial x_{0j}\partial x_{0k}}, \tag{6.30}$$

subject to the following initial and boundary conditions

$$\mathcal{R}_{\mathcal{S}}(t_0,t_0,\mathbf{x}_0)=1,\ \ \mathbf{x}_0\in\mathcal{S}, \tag{6.31}$$

$$\mathcal{R}_{\mathcal{S}}(t,t_0,\mathbf{x}_0)=0,\ \ \mathbf{x}_0\in\Gamma. \tag{6.32}$$

When $\mathbf{X}(t)$ is a stationary process such that $p_{\mathbf{X}}(\mathbf{x},t|\mathbf{x}_0,t_0)=p_{\mathbf{X}}(\mathbf{x},t-t_0|\mathbf{x}_0)$, $A_j(\mathbf{x}_0,t_0)=A_j(\mathbf{x}_0)$ and $B_{jk}(\mathbf{x}_0,t_0)=B_{jk}(\mathbf{x}_0)$. Equation (6.29) implies that $\mathcal{R}_{\mathcal{S}}(t,t_0,\mathbf{x}_0)=\mathcal{R}_{\mathcal{S}}(t-t_0,\mathbf{x}_0)$. Let $\tau=t-t_0$. Equation (6.30) becomes

$$\frac{\partial \mathcal{R}_{\mathcal{S}}(\tau, \mathbf{x}_0)}{\partial \tau} = \sum_{j=1}^{n} A_j(\mathbf{x}_0) \frac{\partial \mathcal{R}_{\mathcal{S}}(\tau, \mathbf{x}_0)}{\partial x_{0j}} + \sum_{j=1}^{n}\sum_{k=1}^{n} \frac{B_{jk}(\mathbf{x}_0)}{2!} \frac{\partial^2 \mathcal{R}_{\mathcal{S}}(\tau, \mathbf{x}_0)}{\partial x_{0j} \partial x_{0k}}. \tag{6.33}$$

The initial and boundary conditions are

$$\mathcal{R}_{\mathcal{S}}(0, \mathbf{x}_0) = 1, \;\; \mathbf{x}_0 \in \mathcal{S}, \tag{6.34}$$

$$\mathcal{R}_{\mathcal{S}}(\tau, \mathbf{x}_0) = 0, \;\; \mathbf{x}_0 \in \Gamma. \tag{6.35}$$

6.4 First-Passage Time Probability

Denote the complement of $\mathcal{R}_{\mathcal{S}}(t, t_0, \mathbf{x}_0)$ as $F_{\mathcal{S}}(t, t_0, \mathbf{x}_0)$, which is the probability distribution function of the first-passage time. We have

$$F_{\mathcal{S}}(t, t_0, \mathbf{x}_0) = P(t \geq T | \mathbf{X}(t_0) = \mathbf{x}_0) = 1 - \mathcal{R}_{\mathcal{S}}(t, t_0, \mathbf{x}_0). \tag{6.36}$$

Substituting this relationship to Eq. (6.30), we obtain

$$-\frac{\partial F_{\mathcal{S}}(t, t_0, \mathbf{x}_0)}{\partial t_0} = \sum_{j=1}^{n} A_j(\mathbf{x}_0, t_0) \frac{\partial F_{\mathcal{S}}(t, t_0, \mathbf{x}_0)}{\partial x_{0j}} + \sum_{j=1}^{n}\sum_{k=1}^{n} \frac{B_{jk}(\mathbf{x}_0, t_0)}{2!} \frac{\partial^2 F_{\mathcal{S}}(t, t_0, \mathbf{x}_0)}{\partial x_{0j} \partial x_{0k}}. \tag{6.37}$$

The probability density function of the first-passage time denoted by $p_T(t|\mathbf{x}_0, t_0)$ is given by

$$p_T(t|\mathbf{x}_0, t_0) = \frac{\partial F_{\mathcal{S}}(t, t_0, \mathbf{x}_0)}{\partial t} = -\frac{\partial \mathcal{R}_{\mathcal{S}}(t, t_0, \mathbf{x}_0)}{\partial t}. \tag{6.38}$$

Differentiating Eq. (6.37) with respect to t, we yield the governing equation for $p_T(t|\mathbf{x}_0, t_0)$

$$-\frac{\partial p_T(t|\mathbf{x}_0, t_0)}{\partial t_0} = \sum_{j=1}^{n} A_j(\mathbf{x}_0, t_0) \frac{\partial p_T(t|\mathbf{x}_0, t_0)}{\partial x_{0j}} + \sum_{j=1}^{n}\sum_{k=1}^{n} \frac{B_{jk}(\mathbf{x}_0, t_0)}{2!} \frac{\partial^2 p_T(t|\mathbf{x}_0, t_0)}{\partial x_{0j} \partial x_{0k}}. \tag{6.39}$$

Since, at a given time $t > t_0$ and when $\mathbf{x}_0 \in \Gamma$, the reliability of the system vanishes $\mathcal{R}_{\mathcal{S}}(t, t_0, \mathbf{x}_0) = 0$, this suggests the boundary condition

$$p_T(t|\mathbf{x}_0, t_0) = 0, \ t > t_0, \ \mathbf{x}_0 \in \Gamma. \tag{6.40}$$

Assume that initially, the system starts from a point in the safe domain with probability one, we have an initial condition

$$p_T(t_0|\mathbf{x}_0, t_0) = \delta(\mathbf{x}_0), \ \mathbf{x}_0 \in \mathcal{S}. \tag{6.41}$$

The first-passage time is a random variable and its rth order moment can be defined as

$$M_r(\mathbf{x}_0, t_0) = \mathrm{E}[(T - t_0)^r|\mathbf{x}_0, t_0] = \int_{t_0}^{\infty} (t - t_0)^r p_T(t|\mathbf{x}_0, t_0)\, dt. \tag{6.42}$$

From Eq. (6.39), we obtain a set of integral-partial differential equations for the *moments of the first-passage time*.

The equations for reliability against first-passage failure and the first-passage time probability can be further simplified for stationary response processes. We skip the discussion here.

6.5 Common Safe Domains

In the literature of random vibrations, the following three safe domains for SDOF systems are commonly studied. They are known as the single-sided type B, the double-sided type D and the envelope type E as illustrated in Fig. 6.1. It is noted

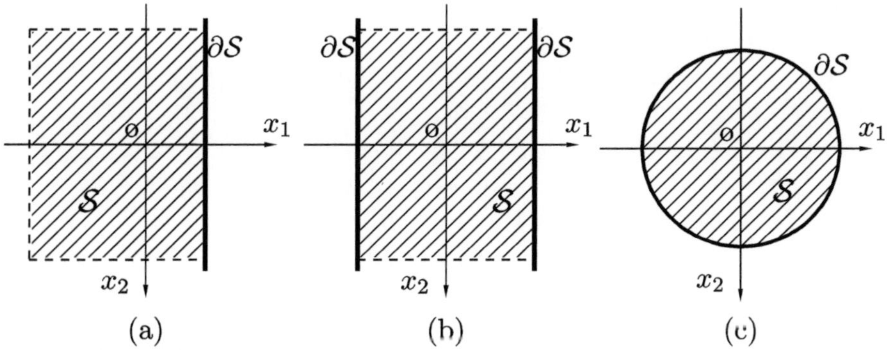

Fig. 6.1 Three common safe domains for single degree of stochastic dynamic systems. **a** Type B: Single-sided domain. **b** Type D: Double-sided domain. **c** Type E: Envelope domain

that type B and type D domains are open sets and can be problematic to specify for multi-degree-of-freedom systems. In the computation to deal with open domains, we usually consider a large enough region such that the probability of the system response leaving the computational domain without first crossing the barrier boundary ∂S is nearly zero.

For multiple degree of freedom systems, the safe domain is more complex. Energy is usually a quantity that can be used to define the safe domain.

References

1. Lin YK (1967) Probabilistic theory of structural dynamics. McGraw-Hill, New York
2. Risken H (1984) The Fokker-Planck equation - methods of solution and applications. Springer, New York
3. Soize C (1994) The Fokker-Planck equation for stochastic dynamical systems and its explicit steady state solutions. World Scientific Publishing Company, River Edge, New Jersey

Chapter 7
Classical Random Vibration

Random vibration is a subject of response study of dynamic systems subject to random excitations. The topics of random vibration analysis includes time domain solution of response and response moments, power spectral of the response, probability distribution and reliability of the system. For linear and nonlinear random vibration of discrete and continuous systems, there are various methods available [1–10].

In this chapter, we present a summary of the classical studies of linear and nonlinear random vibration.

7.1 Linear Random Vibration

We use a SDOF spring-mass oscillator subject to an external random loading $W(t)$ to illustrate the discussion. The motion of the system is governed by the stochastic differential equation

$$\begin{aligned} &m\ddot{X} + c\dot{X} + kX = W(t), \ t > 0, \\ &X(0) = X_0, \ \dot{X}(0) = \dot{X}_0, \end{aligned} \tag{7.1}$$

where m is the mass, k is the spring constant, and c is the damping coefficient. Let $h(t)$ be the impulse response of the system

$$h(t) = \begin{cases} \frac{e^{-\zeta\omega_n t}}{m\omega_d} \sin(\omega_d t), & t \geq 0, \\ 0, & t < 0, \end{cases} \tag{7.2}$$

where $\omega_n = \sqrt{\frac{k}{m}}$ is the *resonant frequency*, $\zeta = \frac{c}{2\omega_n m}$ is the *damping ratio* and $\omega_d = \omega_n\sqrt{1-\zeta^2}$ is the *damped resonant frequency*.

J.-Q. Sun, *Random Vibration with Machine Learning Method*,
Nonlinear Physical Science, https://doi.org/10.1007/978-981-95-0812-9_7

The response of the system is given by *Duhamel's integral* in the mean square sense

$$X(t) = X_h(t) + \int_0^t h(t-\tau)W(\tau)d\tau, \tag{7.3}$$

where $X_h(t)$ is the homogeneous solution of the equation and the response due to initial conditions

$$X_h(t) = (h(t)c + \dot{h}(t)m)X_0 + h(t)m\dot{X}_0. \tag{7.4}$$

With the help of the convolution solution in Eq. (7.3), all other response statistics can be obtained including mean and autocovariance functions of the response $X(t)$

$$\mu_X(t) = \mathrm{E}[X(t)] = \mu_{X_h}(t) + \int_0^t h(t-\tau)\mu_W(\tau)d\tau, \tag{7.5}$$

$$\begin{aligned}\kappa_{XX}(t_1,t_2) &= \mathrm{E}[(X(t_1)-\mu_X(t_1))(X(t_2)-\mu_X(t_2))] \\ &= \kappa_{X_hX_h}(t_1,t_2) \\ &+ \int_0^{t_1}\int_0^{t_2} h(t_1-\tau_1)h(t_2-\tau_2)\kappa_{WW}(\tau_1,\tau_2)d\tau_1 d\tau_2.\end{aligned} \tag{7.6}$$

When the excitation $W(t)$ is a weakly or strictly stationary random process such that its mean is constant and $\kappa_{WW}(\tau_1,\tau_2) = \kappa_{WW}(\tau_2-\tau_1)$, we have

$$\mu_X = \lim_{t\to\infty}\mu_X(t) = \mu_W\int_0^\infty h(\tau)d\tau = \frac{1}{k}\mu_W. \tag{7.7}$$

$$\begin{aligned}\kappa_{XX}(t_1,t_2) &= \int_0^\infty\int_0^\infty h(\tau_1)h(\tau_2)\kappa_{WW}(\tau-\tau_1+\tau_2)d\tau_1 d\tau_2 \\ &= \kappa_{XX}(\tau).\end{aligned} \tag{7.8}$$

Note that $\kappa_{X_hX_h}(t_1,t_2) \to 0$ as $t_1, t_2 \to \infty$.

Example 7.1 Consider a white noise excitation such that $\mu_W = 0$ and $\kappa_{WW}(t_2 - t_1) = 2\pi S_0\delta(t_2 - t_1)$. Assume that the initial conditions of the system are all zero so that $X_h(t) = 0$. Then, for $t_1, t_2 > 0$,

$$
\begin{aligned}
&\kappa_{XX}(t_1, t_2) \qquad (7.9)\\
&= 2\pi S_0 \int_0^{t_1}\int_0^{t_2} h(\tau_1)h(\tau_2)\delta(t_2 - t_1 - \tau_1 + \tau_2)d\tau_1 d\tau_2\\
&= \frac{\pi S_0}{2m^2\zeta\omega_n^3}\left\{e^{-\zeta\omega_n|t_2-t_1|}\left[\cos\omega_d(t_1-t_2) + \frac{\zeta\omega_n}{\omega_d}\sin\omega_d|t_2-t_1|\right]\right.\\
&\quad - e^{-\zeta\omega_n(t_1+t_2)}\left[\frac{\omega_n^2}{\omega_d^2}\cos\omega_d(t_2-t_1) + \frac{\zeta\omega_n}{\omega_d}\sin\omega_d(t_1+t_2)\right.\\
&\quad \left.\left. - \frac{\zeta^2\omega_n^2}{\omega_d^2}\cos\omega_d(t_1+t_2)\right]\right\}.
\end{aligned}
$$

As $t_1, t_2 \to \infty$ and $\tau = t_2 - t_1$ is finite, we have

$$
\begin{aligned}
\kappa_{XX}(t_1, t_2) &= \frac{\pi S_0}{2\zeta\omega_n^3 m^2} e^{-\zeta\omega_n|\tau|}\left[\cos\omega_d\tau + \frac{\zeta\omega_n}{\omega_d}\sin\omega_d|\tau|\right]\\
&= \kappa_{XX}(\tau), \qquad (7.10)
\end{aligned}
$$

and the steady state mean square response

$$
\sigma_X^2 = \kappa_{XX}(0) = \frac{\pi S_0}{2\zeta\omega_n^3 m^2}. \qquad (7.11)
$$

From the correlation function $\kappa_{XX}(\tau) = R_{XX}(\tau)$, we can compute the power spectral density function as

$$
\begin{aligned}
S_{XX}(\omega) &= \frac{1}{2\pi}\int_{-\infty}^{\infty} \kappa_{XX}(\tau)e^{-i\omega\tau}d\tau \qquad (7.12)\\
&= \frac{1}{2\pi}\int_{-\infty}^{\infty}\int_{-\infty}^{\infty}\int_{-\infty}^{\infty} h(\tau_1)h(\tau_2)\\
&\quad \times \kappa_{WW}(\tau - \tau_1 + \tau_2)e^{-i\omega\tau}d\tau_1 d\tau_2 d\tau.
\end{aligned}
$$

The Fourier transformation of the impulse response is the frequency response function.

$$
h(t) = \frac{1}{2\pi}\int_{-\infty}^{\infty} H(\omega)e^{-i\omega t}d\omega, \qquad (7.13)
$$

where

$$H(\omega) = \int_{-\infty}^{\infty} h(\tau)e^{i\omega\tau}d\tau. \tag{7.14}$$

Also, we have

$$\kappa_{WW}(\tau) = \int_{-\infty}^{\infty} S_{WW}(\omega)e^{i\omega\tau}d\omega. \tag{7.15}$$

Substituting all these relationships to Eq. (7.12), we have

$$S_{XX}(\omega) = |H(\omega)|^2 S_{WW}(\omega). \tag{7.16}$$

The stationary mean square response of $X(t)$ can be expressed in terms of $S_{XX}(\omega)$.

The power spectra of derivative processes of various order can be defined in a similar way. We shall skip the discussion. The reader can refer to [9].

Linear random vibration analysis can also be done in the state space with the help of Itô differential equation. For example, the equation of motion (7.1) can be written in the Itô sense as

$$\begin{aligned} dX_1(t) &= X_2 dt, \quad t > 0, \ , \\ dX_2(t) &= \left[-\omega_n^2 X_1 - 2\zeta\omega_n X_2\right] dt + \sigma_{21} dB_1(t), \\ \mathbf{X}(0) &= \mathbf{x}_0, \end{aligned} \tag{7.17}$$

where

$$\begin{aligned} m_1(\mathbf{X}) &= X_2, \ m_2(\mathbf{X}) = -\omega_n^2 X_1 - 2\zeta\omega_n X_2, \\ \sigma_{11} &= 0, \ \sigma_{21} = \frac{1}{m}\sqrt{2\pi S_0}, \end{aligned} \tag{7.18}$$

and the unit Brownian motion $B_1(t)$ has the following properties

$$\mathrm{E}[dB_1(t)] = 0, \ \mathrm{E}[dB_1(t)dB_1(s)] = \begin{cases} dt, \ t = s, \\ 0, \ t \neq s. \end{cases} \tag{7.19}$$

In the state space form, it is convenient to derive equations governing the response moments. Consider the derivative of a polynomial function $X_1^k X_2^l$ by applying Itô's lemma. We have

$$\begin{aligned} d(X_1^k X_2^l) = & [m_1(\mathbf{X})kX_1^{k-1}X_2^l + m_2(\mathbf{X})lX_1^k X_2^{l-1}]dt \\ & + \frac{\sigma_{21}^2}{2} l(l-1)X_1^k X_2^{l-2} dt \\ & + \sigma_{21} l X_1^k X_2^{l-1} dB_1(t) + O(dt^2). \end{aligned} \tag{7.20}$$

By selecting the pairs of the power indices (k, l) to be (1, 0), (0, 1), (2, 0), (1, 1), and (0, 2), we arrive at a set of moment equations

$$\begin{aligned}\frac{d\mathrm{E}[X_1]}{dt} &= \mathrm{E}[X_2],\\ \frac{d\mathrm{E}[X_2]}{dt} &= -\omega_n^2\mathrm{E}[X_1] - 2\zeta\omega_n\mathrm{E}[X_2],\end{aligned} \tag{7.21}$$

and

$$\begin{aligned}\frac{d\mathrm{E}[X_1^2]}{dt} &= 2\mathrm{E}[X_1X_2],\\ \frac{d\mathrm{E}[X_1X_2]}{dt} &= \mathrm{E}[X_2^2] - \omega_n^2\mathrm{E}[X_1^2] - 2\zeta\omega_n\mathrm{E}[X_1X_2],\\ \frac{d\mathrm{E}[X_2^2]}{dt} &= -2\omega_n^2\mathrm{E}[X_1X_2] - 4\zeta\omega_n\mathrm{E}[X_2^2] + \frac{2\pi S_0}{m^2}.\end{aligned} \tag{7.22}$$

The higher order moment equations can be constructed in the same fashion. The moment equations are deterministic ordinary differential equations and satisfy the initial conditions.

Assume that the system starts at a point $\mathbf{x}_0$ with probability one. Then, the initial conditions for the moments are

$$\begin{aligned}\mathrm{E}[X_1] &= x_{10},\ \ \mathrm{E}[X_2] = x_{20},\\ \mathrm{E}[X_1X_2] &= x_{10}x_{20},\\ \mathrm{E}[X_1^2] &= x_{10}^2,\ \ \mathrm{E}[X_2^2] = x_{20}^2,\ \ \text{at } t = 0.\end{aligned} \tag{7.23}$$

The steady state solutions of the moment equations are obtained by setting the time derivatives of the moments to zero in Eqs. (7.21) and (7.22) leading to a set of algebraic equations.

To obtain the probability distribution and other statistics of the response of the oscillator, we need to obtain the probability density function from the Fokker-Planck-Kolmogorov (FPK) equation.

The FPK equation for the linear oscillator reads,

$$\begin{aligned}\frac{\partial}{\partial t}p_{\mathbf{X}}(\mathbf{x}, t|\mathbf{x}_0, t_0) = &-\frac{\partial}{\partial x_1}[x_2 p_{\mathbf{X}}]\\ &-\frac{\partial}{\partial x_2}[(-\omega_n^2 x_1 - 2\zeta\omega_n x_2)p_{\mathbf{X}}]\\ &+\frac{\pi S_0}{m^2}\frac{\partial^2 p_{\mathbf{X}}}{\partial x_2^2}.\end{aligned} \tag{7.24}$$

Let the initial condition be given by

$$p_{\mathbf{X}}(\mathbf{x}, t_0|\mathbf{x}_0, t_0) = \delta(\mathbf{x} - \mathbf{x}_0). \tag{7.25}$$

$p_{\mathbf{X}}(\mathbf{x}, t|\mathbf{x}_0, t_0)$ is obtained as a Gaussian probability density function with mean and variance matrix given by

$$\mu_{\mathbf{X}}(t) = \begin{bmatrix} (h(t)c + \dot{h}(t)m)x_0 + h(t)m\dot{x}_0 \\ (\dot{h}(t)c + \ddot{h}(t)m)x_0 + \dot{h}(t)m\dot{x}_0 \end{bmatrix}, \tag{7.26}$$

and

$$\mathbf{C}_{\mathbf{XX}}(t) = \begin{bmatrix} \kappa_{XX}(t) & \kappa_{X\dot{X}}(t) \\ \kappa_{X\dot{X}}(t) & \kappa_{\dot{X}\dot{X}}(t) \end{bmatrix}, \tag{7.27}$$

where

$$\begin{aligned} \kappa_{XX}(t) &= \kappa_{XX}(t_1, t_2)|_{t_1, t_2 = t}, \\ \kappa_{X\dot{X}}(t) &= \left.\frac{\partial \kappa_{XX}(t_1, t_2)}{\partial t_2}\right|_{t_1, t_2 = t}, \\ \kappa_{\dot{X}\dot{X}}(t) &= \left.\frac{\partial^2 \kappa_{XX}(t_1, t_2)}{\partial t_1 \partial t_2}\right|_{t_1, t_2 = t}. \end{aligned} \tag{7.28}$$

$\kappa_{XX}(t_1, t_2)$ is given in Eq. (7.9).

Except for the reliability problem, the above discussions demonstrate the complete solution process of linear random vibration analysis.

Next, we consider nonlinear random vibration problems with the classical approach.

7.2 Nonlinear Random Vibration

For nonlinear dynamic systems, it is usually difficult or impossible to obtain time-domain random vibration response analytically. Hence, there is a shortage of studies on power spectral density of random vibration.

Analytical methods such as equivalent linearization, cumulant neglect and stochastic averaging can provide approximate solutions of the response when the system is weakly nonlinear. Monte Carlo simulation of random vibration of non-linear dynamic systems is the only universal method for numerical solutions and is known to be time-consuming.

This section presents several popular methods for the solution of transient and steady-state solutions of the FPK equation for nonlinear random vibration.

7.2.1 Short-Time Solution

As an example, we consider a one-dimensional FPK equation as an example. We rewrite the equation in *operator form* as

$$\frac{\partial}{\partial t} p_X(x, t|x_0, t_0) = \mathcal{L}_{FPK}(x, t) p_X, \tag{7.29}$$

where

$$\mathcal{L}_{FPK}(x, t) = -\frac{\partial}{\partial x} A(x, t) + \frac{1}{2}\frac{\partial^2}{\partial x^2} B(x, t), \tag{7.30}$$

subject to the initial condition

$$p_X(x, t|x_0, t_0) = \delta(x - x_0). \tag{7.31}$$

The formal solution of Eq. (7.29) can be written as [11]

$$p_X(x, t|x_0, t_0) = e^{\mathcal{L}_{FPK}(x,t)\cdot(t-t_0)} \delta(x - x_0). \tag{7.32}$$

Let $\tau = t - t_0$ and assume that $\tau << 1$. Expand the solution with respect to τ. We have the short-time solution

$$p_X(x, t_0 + \tau|x_0, t_0) = [1 + \mathcal{L}_{FPK}(x, t) \cdot \tau + O(\tau^2)]\delta(x - x_0). \tag{7.33}$$

The following identity in the generalized function sense is applied to the short-time solution

$$\delta(x - x_0) f(x) = \delta(x - x_0) f(x_0), \tag{7.34}$$

leading to

$$p_X(x, t_0 + \tau|x_0, t_0) = e^{-\frac{\partial}{\partial x} A(x_0,t)\tau + \frac{1}{2}\frac{\partial^2}{\partial x^2} B(x_0,t)\tau} \delta(x - x_0). \tag{7.35}$$

Substituting the generalized function representation of $\delta(x - x_0)$

$$\delta(x - x_0) = \frac{1}{2\pi} \int_{-\infty}^{\infty} e^{i\theta(x-x_0)} d\theta, \tag{7.36}$$

to this solution, we obtain

$$p_X(x, t_0 + \tau|x_0, t_0) \tag{7.37}$$

$$
= \frac{1}{2\pi} \int_{-\infty}^{\infty} \exp\left[-\frac{\partial}{\partial x} A(x_0, t)\tau + \frac{1}{2} \frac{\partial^2}{\partial x^2} B(x_0, t)\tau + i\theta(x - x_0) \right] d\theta
$$

$$
= \frac{1}{2\pi} \int_{-\infty}^{\infty} \exp\left[-i\theta A(x_0, t)\tau - \frac{\theta^2}{2} B(x_0, t)\tau + i\theta(x - x_0) \right] d\theta
$$

$$
= \frac{1}{\sqrt{2\pi \cdot 2B(x_0, t)\tau}} \exp\left[-\frac{[x - x_0 - A(x_0, t)\tau]^2}{4B(x_0, t)\tau} \right].
$$

Hence, the short-time solution of the FPK equation starting from the initial condition $\delta(x - x_0)$ is a Gaussian probability density function with the mean $x_0 + A(x_0, t)\tau$ and variance $2B(x_0, t)\tau$.

7.2.2 Improvement of Short-Time Solution

The short-time solution presented above is based on the Taylor expansion of the first order accuracy in τ. If the short-time solution is viewed as a numerical integration of the FPK equation in time domain, it is comparable to the Euler integration rule. To improve the accuracy of the short-time solution while keeping the normality of the short-time solution, we naturally look for higher order estimates of the mean and variance.

One approach for nonlinear systems is to apply the Gaussian closure or non-Gaussian closure to the moment equations of the system response. This allows for more accurate estimates of the mean and variance over a longer time period, leading to more accurate Gaussian probability density function over somewhat longer time interval [12, 13].

7.2.3 Path Integral

Let $p_X(x_0, t_0)$ denote the initial probability density function of the stochastic process $X(t)$. Recall that

$$
p_X(x, t_0 + \tau) = \int_{R^1} p_X(x, t_0 + \tau | x_0, t_0) p_X(x_0, t_0) dx_0. \tag{7.38}
$$

Repeating this step over the time interval $t_0 + \tau$ to $t_0 + 2\tau$, and applying the short-time solution leading to a general expression over n steps, we obtain

$$p_X(x, t_0 + n\tau) = \int_{R^1}\int_{R^1} \cdots \int_{R^1} \prod_{k=0}^{n-1} \frac{p_X(x_0, t_0)}{\sqrt{4\pi B(x_k, t_0 + k\tau)\tau}} \times \exp\left[-\frac{[x_{k+1} - x_k - A(x_k, t_0 + k\tau)\tau]^2}{4B(x_k, t_0 + k\tau)\tau}\right] dx_k. \tag{7.39}$$

This is known as the *path integral of the FPK equation*. The path integral converges to the correct solution at a given $t > t_0$ as $n \to \infty$ with the accuracy of the order $1/n$ [11].

7.2.4 Markov Chain Representation

To compute the path integral, we have to discretize the solution. Let us partition R^1 into equal-intervals with the jth interval defined as $x_{j-1} \le x < x_j$. Define a probability vector as

$$p_j(k) = \int_{x_{j-1}}^{x_j} p_X(x, t_0 + k\tau)dx. \tag{7.40}$$

$p_j(k)$ is the probability that the system lies in the interval $x_{j-1} \le x < x_j$ at the kth time step. We further assume that this probability is concentrated in the center $x_{j-1/2} = (x_j + x_{j-1})/2$ of the interval. Hence, we have an approximate histogram representation of $p_X(x, t_0 + k\tau)$ as follows

$$p_X(x, t_0 + k\tau) = \sum_j \frac{p_j(k)}{\Delta x_j}\delta(x - x_{j-1/2}), \tag{7.41}$$

where $\Delta x_j = x_j - x_{j-1}$.

The path integral solution in Eq. (7.39) is expressed in the discrete space as

$$p_j(k) = \sum_{l_k, l_{k-1}, \cdots, l_0} P_{jl_k}(k) P_{l_k l_{k-1}}(k-1) \cdots P_{l_1 l_0}(1) p_{l_0}(0), \tag{7.42}$$

where

$$P_{jl}(k) = \int_{x_{j-1}}^{x_j}\int_{x_{l-1}}^{x_l} p_X(x, t_0 + k\tau | x_1, t_0 + (k-1)\tau) dx dx_1 \times \frac{1}{\Delta x_j \Delta x_l}. \tag{7.43}$$

$P_{jl}(k)$ is the one step transition probability at the kth time step from the lth interval to the jth interval. Introduce the vector notations: $\mathbf{p}(k) = \{p_j(k)\}$ and $\mathbf{P}(k) = \{P_{jl}(k)\}$. Equation (7.42) now reads

$$\mathbf{p}(k) = \mathbf{P}(k)\mathbf{P}(k-1)\cdots\mathbf{P}(1)\mathbf{p}(0). \tag{7.44}$$

This equation defines a nonstationary Markov chain.

When the short-time solution is substituted in Eqs. (7.43), (7.44) becomes the Markov chain representation of the path integral solution. This discrete representation is readily amenable to numerical computations.

When the stochastic process $X(t)$ is weakly stationary such that $p_X(x, t_0 + k\tau|x_1, t_0 + (k-1)\tau) = p_X(x, t_0 + \tau|x_1, t_0)$ and $P_{jl}(k)$ is independent of k, we have

$$\mathbf{P}(k) = \mathbf{P}(k-1) = \cdots = \mathbf{P}(1) \equiv \mathbf{P}. \tag{7.45}$$

Equation (7.44) is reduced to

$$\mathbf{p}(k) = \mathbf{P}^k\mathbf{p}(0). \tag{7.46}$$

The equation represents a stationary Markov chain.

This solution is also known as the short-time Gaussian generalized cell mapping, and can be applied to a vector process $\mathbf{X}(t)$ [14]. There have been many studies of random vibration problems with the generalized cell mapping method [13–20].

7.2.5 Exact Solution of Stationary PDF

When the stochastic differential equation is nonlinear, it is in general very difficult to obtain *exact* transient or stationary solutions of the FPK equation. Under certain conditions, we can, however, obtain exact stationary solution of the FPK equation. Here, we list the known examples of exact solutions of stationary probability density function for the FPK equation.

$$\begin{aligned} &-\sum_{k=1}^{n} \frac{\partial}{\partial x_k}[A_k(\mathbf{x}, t)p_{\mathbf{X}}] \\ &+\sum_{j=1}^{n}\sum_{k=1}^{n} \frac{\partial^2}{\partial x_j \partial x_k}\left[\frac{B_{jk}(\mathbf{x}, t)}{2!}p_{\mathbf{X}}\right] = 0. \end{aligned} \tag{7.47}$$

1. The stationary probability density function of the first order system, if it exists, is independent of time, and satisfies the reduced FPK equation

$$-\frac{d}{dx}[A(x)p_X(x)] + \frac{1}{2}\frac{d^2}{dx^2}[B(x)p_X(x)] = 0. \tag{7.48}$$

When the system is defined in R^1 and stable in the sense that as $|x| \to \infty$, $p_X(x) \to 0$ and $\frac{d}{dx} p_X \to 0$, we obtain

$$p_X(x) = \frac{C_0}{B(x)} \exp\left[2 \int_0^x \frac{A(z)}{B(z)} dz\right]. \tag{7.49}$$

where C_0 is a normalization constant.

2. Consider a general second order nonlinear system known as Yasuda's system [21],

$$\ddot{x} + g(x, \dot{x}) = \eta_1 W_1(t) + \eta_2 x W_2(t), \tag{7.50}$$

where $W_j(t)$ are independent Gaussian white noises such that

$$\begin{aligned} \mathrm{E}[W_j(t)] &= 0, \\ \mathrm{E}[W_j(t) W_k(t+\tau)] &= 2 D_k \delta_{jk} \delta(\tau). \end{aligned} \tag{7.51}$$

It is noted that the physical equation of motion is to be interpreted in the Strotonovich when it is converted the state equation. The Wong-Zakai correction term $\frac{1}{2} \sum_{k=1}^{n} \sum_{l=1}^{m} \frac{\partial \sigma_{jl}(\mathbf{X},t)}{\partial X_k} \sigma_{kl}(\mathbf{X}, t)$ is zero for this system. Hence, the equation in the Itô sense is the same as the equation in the Strotonovich sense. The drift and diffusion coefficients are given by

$$\begin{aligned} &m_1 = x_2, \quad m_2 = -g(x_1, x_2), \\ &[b_{jk}] = \left[\sum_{l=1}^{m} \sigma_{kl} \sigma_{jl}\right] = \begin{bmatrix} 0 & 0 \\ 0 & 2\eta_1^2 D_1 + 2\eta_2^2 D_2 x_1^2 \end{bmatrix}. \end{aligned} \tag{7.52}$$

The stationary probability density function $p_{\mathbf{X}}(\mathbf{x})$ satisfies the reduced FPK equation

$$\begin{aligned} &\left[\eta_1^2 D_1 + \eta_2^2 D_2 x_1^2\right] \frac{\partial^2 p_{\mathbf{X}}(x_1, x_2)}{\partial x_2^2} \\ &- \frac{\partial}{\partial x_1}[x_2 p_{\mathbf{X}}] + \frac{\partial}{\partial x_2}[g(x_1, x_2) p_{\mathbf{X}}] = 0. \end{aligned} \tag{7.53}$$

Assume that the solution is in the following form

$$p_{\mathbf{X}}(x_1, x_2) = C_0 e^{f(x_1, x_2)}, \tag{7.54}$$

where C_0 is the normalization constant and $f(x_1, x_2)$ is a function that can be analytically obtained from the reduced FPK equation under certain conditions [9].

3. Dimentberg's system is another interesting example that has a closed-form solution to the reduced FPK equation.
The nonlinear Stratonovich stochastic equation reads [22],

$$\begin{aligned} &\ddot{x} + 2\alpha\dot{x}[1+\eta(t)] + \beta_1 \dot{x}\left[x^2 + \frac{\dot{x}^2}{\Omega^2}\right] \\ &+ \Omega^2 x[1+\xi(t)] = \zeta(t), \end{aligned} \tag{7.55}$$

where $\beta_1 \geq 0$, $\alpha > 0$, $\eta(t)$, $\xi(t)$ and $\zeta(t)$ are zero-mean independent Gaussian physical white noises with the correlation functions given by

$$\begin{aligned} \mathrm{E}[\eta(t)\eta(t+\tau)] &= D_\eta \delta(\tau), \\ \mathrm{E}[\xi(t)\xi(t+\tau)] &= D_\xi \delta(\tau), \\ \mathrm{E}[\zeta(t)\zeta(t+\tau)] &= D_\zeta \delta(\tau). \end{aligned} \tag{7.56}$$

The stationary probability density function $p_{\mathbf{X}}(\mathbf{x})$ satisfies the reduced FPK equation

$$\begin{aligned} 0 = &\frac{1}{2}\frac{\partial^2}{\partial x_2^2}\left\{\left[4\alpha^2 D_\eta x_2^2 + \Omega^4 D_\xi x_1^2 + D_\zeta\right] p_{\mathbf{X}}(x_1, x_2)\right\} \\ &- \frac{\partial}{\partial x_1}[x_2 p_{\mathbf{X}}] \\ &+ \frac{\partial}{\partial x_2}\left\{\left[2\alpha x_2 + \beta_1 x_2\left[x_1^2 + \frac{x_2^2}{\Omega^2}\right] + \Omega^2 x_1 - 2\alpha^2 D_\eta x_2\right] p_{\mathbf{X}}\right\}. \end{aligned} \tag{7.57}$$

If

$$\Omega^2 D_\xi = 4\alpha^2 D_\eta, \tag{7.58}$$

the reduced FPK equation has an exact solution

$$p_{\mathbf{X}}(x_1, x_2) = C_0 \frac{\exp\left[-\beta\left(x_1^2 + \frac{x_2^2}{\Omega^2}\right)\right]}{\left(\kappa + x_1^2 + \frac{x_2^2}{\Omega^2}\right)^{\delta - \kappa\beta}}, \tag{7.59}$$

where

$$\kappa = \frac{D_\zeta}{\Omega^4 D_\xi}, \quad \delta = \frac{2\alpha}{\Omega^2 D_\xi} + \frac{1}{2}, \quad \beta = \frac{\beta_1}{\Omega^2 D_\xi}, \tag{7.60}$$

and C_0 is the normalization constant.

4. Many Hamiltonian systems also allow the closed-form solution of the reduced FPK equation [23].

7.2.6 *Detailed Balance*

The method of detailed balance provides a systematic way to find closed-form solutions of the reduced FPK equation for nonlinear stochastic systems with invertible or non-invertible diffusion matrix. The introduction of the method as well as examples of application can be found in [24], [9].

7.3 Equivalent Itô Equation

When an Itô equation is subject to more than one independent Brownian motion excitations, an equivalent Itô equation with only one Brownian motion excitation can be found in the sense that both Itô equations share the same FPK equation.

This is an important result. When we study numerical solutions of a stochastic system subject to more than one independent random excitations by using Monte Carlo simulation, we must simulate all the independent white noise processes. This can be computationally demanding. If we use the equivalent Itô equation, we only need to simulate one Brownian motion. This is far more efficient computationally.

References

1. Elishakoff I (1983) Probabilistic methods in the theory of structures. Wiley, New York
2. Lin YK (1967) Probabilistic theory of structural dynamics. McGraw-Hill, New York
3. Newland DE (1984) Random vibrations and spectral analysis. Longman, New York
4. Nigam NC (1983) Introduction to random vibrations. The MIT Press, Cambridge, Massachusetts
5. Soong TT, Grigoriu M (1993) Random vibration of mechanical and structural systems. Prentice Hall, Englewood Cliffs, New Jersey
6. Yang CY (1986) Random vibration of structures. Wiley, New York
7. Wirsching PH, Paez TL, Ortiz H (1995) Random vibrations - theory and practice. Wiley, New York
8. Sólnes J (1997) Stochastic processes and random vibrations - theory and practice. Wiley, New York
9. Sun JQ (2006) Stochastic dynamics and control. Elsevier Science Ltd., Oxford, U.K
10. Cai G, Zhu WQ (2017) Elements of stochastic dynamics. World Scientific, Hackensack, New Jersey
11. Risken H (1984) The Fokker-Planck equation - methods of solution and applications. Springer, New York
12. Sun JQ, Hsu CS (1989) Cumulant-neglect closure methods for asymmetric non-linear systems driven by Gaussian white noise. J Sound Vib 135(2):338–345
13. Sun JQ, Hsu CS (1990) The generalized cell mapping method in nonlinear random vibration based upon short-time Gaussian approximation. J Appl Mech 57(4):1018–1025
14. Sun JQ, Hsu CS (1989) Random vibration of hinged elastic shallow arch. J Sound Vib 132(2):299–315
15. Sun JQ, Hsu CS (1988) First-passage time probability of non-linear stochastic systems by generalized cell mapping method. J Sound Vib 124(2):233–248

16. Sun JQ (1988) Random vibrations of nonlinear systems based upon the cell state space concept. Ph.D. Dissertation, University of California
17. Sun JQ (1995) Random vibration analysis of a nonlinear system with dry friction damping by the short-time Gaussian cell mapping method. J Sound Vib 180(5):785–795
18. Han Q, Xu W, Sun JQ (2016) Stochastic response and bifurcation of periodically driven nonlinear oscillators by the generalized cell mapping method. Phys A-Stat Mech Its Appl 458:115–125
19. Han Q, Xu W, Hu B, Huang DM, Sun JQ (2018) Extinction time of a stochastic predator-prey model by the generalized cell mapping method. Phys-Stat Mech Its Appl 494:351–366
20. Yue X, Xu W, Xu Y, Sun JQ (2019) Non-stationary response of MDOF dynamical systems under combined Gaussian and Poisson white noises by the generalized cell mapping method. Probab Eng Mech 55:102–108
21. Wang R, Yasuda K (1999) Exact stationary response solutions of nine classes of nonlinear stochastic systems under stochastic parametric and external excitations. In: Proceedings of the ASME design engineering technical conferences. Las Vegas, Nevada, pp 1–15
22. Dimentberg MF (1982) An exact solution to a certain non-linear random vibration problem. Int J Non-Linear Mech 17(4):231–236
23. Zhu WQ, Yang Y (1996) Exact stationary solutions of stochastically excited and dissipated integrable Hamiltonian systems. J Appl Mech 63:493–500
24. Soize C (1994) The Fokker-Planck equation for stochastic dynamical systems and its explicit steady state solutions. World Scientific Publishing Company, River Edge, New Jersey

Chapter 8
Random Vibration with Neural Networks

Except for linear systems, analytical solutions of random vibration problems for nonlinear systems are difficult to obtain. Limited examples of nonlinear stochastic systems are presented earlier in the book, for which the stationary PDF solution of the FPK equation can be obtained. Even for these systems, other aspects of random responses such as moments, correlation function and power spectral density are still elusive. Numerical simulations and approximate analytical solutions are still the focus of research.

This chapter introduces machine learning methods to analyze random vibration of nonlinear systems [1]. In particular, we present a radial basis function neural networks solution of the FPK equation. The activation functions consist of Gaussian probability density functions with the means distributed in the state space. This neural networks approximate solution provides a very efficient and accurate way of computing the time-varying and steady state probability density function of nonlinear dynamic systems subject to random excitations, the first-passage time probability of the system and response moments by using the computed PDF [2].

8.1 Neural Networks

Neural networks methods for solving partial differential equations have become a popular topic [1, 3, 4]. The well-known classical Ritz method is also implemented with the help of neural networks [5]. The finite element solutions are obtained with the help of neural networks [6].

There have been many studies of radial basis functions for regression applications and for solving partial differential equations [7]. The authors of Park and Sandberg [8] conclude that the radial-basis-function networks having one hidden layer and the same smoothing factor in each kernel are broad enough for universal approximation.

J.-Q. Sun, *Random Vibration with Machine Learning Method*,
Nonlinear Physical Science, https://doi.org/10.1007/978-981-95-0812-9_8

An excellent review on radial basis function networks for regression and classification can be found in [9].

Neural networks are modeled after the human brain's network of neurons and are designed to process complex patterns in data. A neural network consists of *neurons* (or nodes). These neurons are organized in layers. Each layer is connected to the next through weighted edges.

- **Input Layer**: The first layer where input data enters the network.
- **Hidden Layer(s)**: Intermediate layers that process data through transformations.
- **Output Layer**: The final layer, which produces the network's result.

Let a neuron receive n inputs x_i, generate an output z and have an *activation function* $\phi(y)$. Mathematically, the neuron computes the following

$$z = \phi(y), \tag{8.1}$$
$$y = \sum_{i=1}^{n} w_i x_i + b,$$

where w_i is the weight applied to input x_i, b is the bias, and y is passed through an activation function $\phi(y)$ to produce the output z.

Commonly used activation functions are discussed below.

8.1.1 Sigmoid Function

The sigmoid activation function is defined as

$$\phi(x) = \frac{1}{1+e^{-x}}. \tag{8.2}$$

Sigmoid function maps any real-valued number into the range (0, 1), which is in a probabilistic range. It is useful for binary classification tasks. Furthermore, it is continuously differentiable, which is essential for gradient-based optimization (Fig. 8.1).

The sigmoid activation function has *vanishing gradient* for very large or small input values. This can slow down gradient descent learning during backpropagation. The output of sigmoid activation function is not zero-centered, which can also lead to slow convergence of learning.

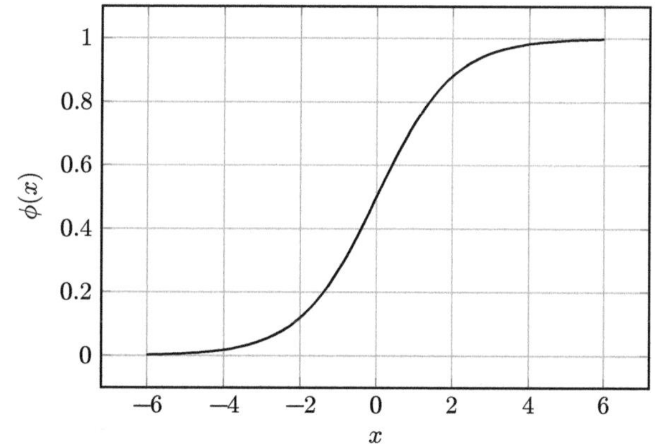

Fig. 8.1 Sigmoid activation function

8.1.2 Hyperbolic Tangent (Tanh)

The hyperbolic tangent function is continuously differentiable and maps input values to the range $(-1, 1)$,

$$\phi(x) = \frac{2}{1 + e^{-2x}} - 1. \tag{8.3}$$

Its output is zero-centered, which helps in faster convergence compared to the sigmoid function. The hyperbolic tangent function is suitable for problems where the output needs to be positive or negative. It also suffers from the *vanishing gradient problem*, similar to the sigmoid function (Fig. 8.2).

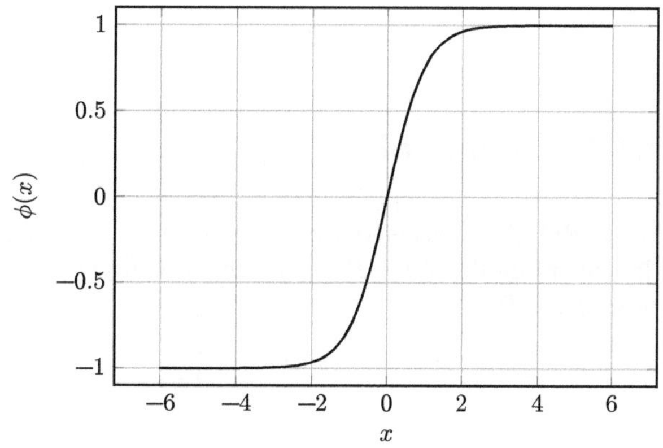

Fig. 8.2 Tanh activation function

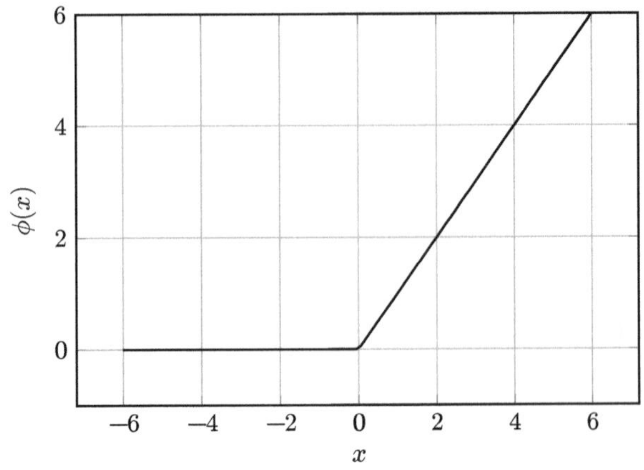

Fig. 8.3 ReLU activation function

8.1.3 Rectified Linear Unit (ReLU)

The ReLU function is defined as

$$\phi(x) = \max(0, x). \tag{8.4}$$

ReLU is computationally efficient, i.e. a thresholding at zero is needed. It helps mitigate the vanishing gradient problem, as it does not saturate for large positive values. ReLU neurons are activated when the input is greater than zero, which introduces sparsity and can improve computational efficiency.

ReLU neurons can become inactive and stop learning entirely when the input stays negative so that the output is always zero. This is known as the *dying ReLU problem* (Fig. 8.3).

8.1.4 Leaky ReLU

Leaky ReLU is a variant of ReLU, which has a small negative slope for negative inputs thus avoiding the *dying ReLU problem* while keeping the computational efficiency of standard ReLU.

$$\phi(x) = \begin{cases} x, & \text{if } x > 0, \\ \alpha x, & \text{if } x \leq 0, \end{cases} \tag{8.5}$$

where α is a small constant (e.g., 0.01). The negative slope α is a hyperparameter that requires tuning (Fig. 8.4).

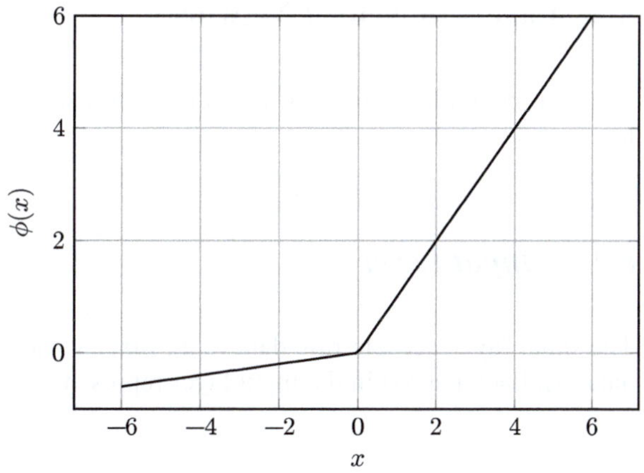

Fig. 8.4 Leaky ReLU activation function

8.1.5 Softmax

The softmax function is typically used in the output layer of a classification network, converting raw scores (logits) into probabilities. The softmax function is susceptible to numerical instability if raw scores (logits) are large (Fig. 8.5).

$$\phi(\mathbf{x})_i = \frac{e^{x_i}}{\sum_j e^{x_j}}. \tag{8.6}$$

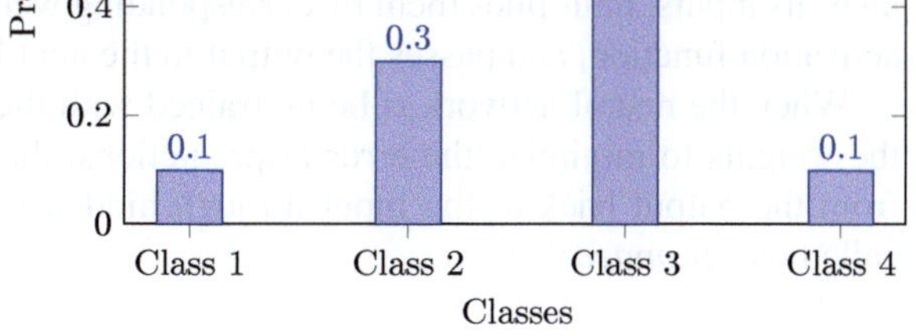

Fig. 8.5 Softmax activation function: output as probabilities

8.2 Layers of Neural Networks

A neural network consists of layers of neurons. These are the input layer, hidden layers and output layer.

8.2.1 Input Layer

The input layer accepts raw data (e.g., pixel values of an image, or features of a dataset). Each neuron in the input layer represents a feature from the data.

8.2.2 Hidden Layer

The hidden layers are between the input and output layers. A neural network can have multiple hidden layers. Each hidden layer can have different number of neurons. The total number of neurons in the hidden layers determines the complexity of the neural network. The hidden layers extract relevant features from the input data by applying transformations and non-linear operations.

8.2.3 Output Layer

The output layer produces the final result, which can be a continuous value, usually for regression, or a class label, commonly for classification. The number of neurons in the output layer is problem dependent. For instance, in binary classification, there is typically one output neuron, while in multi-class classification, there may be one neuron per class.

When a neural network takes the input through hidden layers to predict the output, this process is known as *forward propagation*. Each neuron in the hidden layers takes its inputs, multiplies them by corresponding weights, adds the bias, applies the activation function, and passes the output to the next layer.

When the neural network is being trained with the prediction error by adjusting the weights to minimize the error in predictions, the error information propagates from the output back to the input through hidden layers. This training process is called *backpropagation*.

8.3 Loss Function

The *loss function* measures the accuracy of predictions of the neural network compared to the actual values. Common loss functions include:

- **Mean Squared Error (MSE)** for regression:

$$\text{MSE} = \frac{1}{n}\sum_{i=1}^{n}(y_i - \hat{y}_i)^2. \tag{8.7}$$

- **Cross-Entropy Loss** for classification:

$$L = -\sum y \log(\hat{y}). \tag{8.8}$$

When the neural networks are used to construct the solution of ordinary or partial differential equations, the mean squared errors of the equation are also included as part of the loss. This is because finding the solution of differential equation in terms of neural networks is essentially a more complex or indirect regression problem. It is commonly referred to this class of neural networks as physics informed neural networks (PINN).

Training of the neural networks by backpropagation involves the following steps. The objective is to minimize the loss function by updating the weights iteratively.

1. Compute the loss with the network output and the labeled data.
2. Calculate the gradient of the loss with respect to each trainable parameter or weight of the model using the chain rule of calculus.
3. Update weights using gradient descent or other optimization algorithms.

8.4 Architecture of Neural Networks

Researchers have invented many neural networks that have different architecture and provide diverse functions. Some common ones include Feedforward Neural Networks (FNN), Convolutional Neural Networks (CNN), Recurrent Neural Networks (RNN), AutoEncoder, *etc.*. While they are invented for specific applications, new applications and architectures of neural networks are constantly reported in the literature.

It is interesting to point out that the architecture of neural networks determines the functionality.

8.5 Overfitting and Regularization

For a given set of data, we may select an overly complex neural networks to model it, for example, using deep neural networks, which leads to *overfitting*. In this case, the network performs well on training data, but poorly on unseen data. Hence, the neural network model is lack of ability to generalize.

One way to deal with overfitting issue is to apply regularization techniques such as dropout and L_2 regularization. Dropout randomly "turns off" some neurons during training to prevent the model from becoming overly dependent on specific neurons. L_2 regularization adds a penalty to the loss function to keep the weights small, hence avoiding overfitting.

8.6 Radial Basis Function Neural Networks

Radial basis function neural networks (RBFNN) are a class of artificial neural networks that utilize radial basis functions as neurons. The most common choice for the radial basis function is the Gaussian function. RBFNNs are typically used for interpolation and function approximation tasks due to their ability to approximate smooth and even non-smooth functions.

An RBFNN consists of three layers: the input layer, a hidden layer where each neuron computes the activation based on the distance between the input and the neuron center, and the output layer, which typically performs a weighted sum of the hidden layer activations. Each neuron in the hidden layer uses a radial basis function, such as the Gaussian function, which is defined as

$$\phi(x) = \exp\left(-\frac{(x-\mu)^2}{2\sigma^2}\right), \tag{8.9}$$

where μ is the center and σ is the standard deviation.

In the case of Gaussian activation functions, the network responses are localized around their centers, making RBFNNs particularly effective for capturing localized patterns in data or approximating unknown functions. Gaussian neurons are usually distributed in time or spatial domain to capture the response of dynamic systems governed by ordinary or partial differential equation. A one-dimensional case is illustrated below with five Gaussian activation functions distributed along the spatial domain $x = -5$ to $x = 5$ with means at $x = -4, -2, 0, 2, 4$ and a standard deviation $\sigma = 1$ (Fig. 8.6).

Next, we illustrate how to use RBFNN with Gaussian neurons to find the solution of FPK equations for nonlinear stochastic systems.

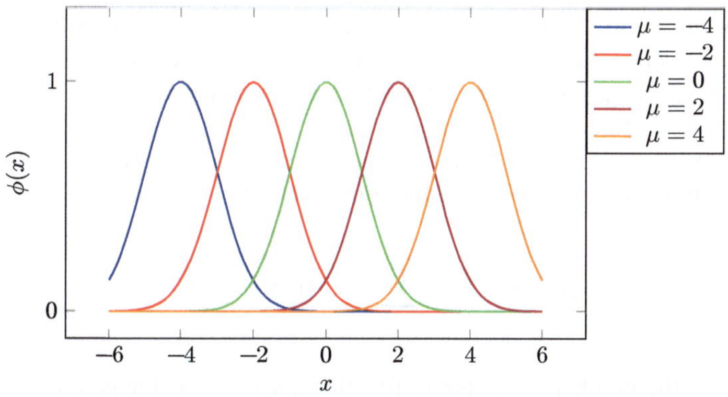

Fig. 8.6 Distributed Gaussian neurons in one dimensional space

8.7 RBFNN Solution of FPK Equation

Consider random vibrations of an n degree of freedom nonlinear system subject to Gaussian white noise excitations. By introducing the state variables, the equation of motion can be written in a $2n$-dimensional state space.

The conditional probability density function $p(\mathbf{x}, t|\mathbf{x}_0, t_0)$ of the system response satisfies the FPK equation

$$
\begin{aligned}
\frac{\partial}{\partial t} p(\mathbf{x}, t|\mathbf{x}_0, t_0) &= -\sum_{k=1}^{2n} \frac{\partial}{\partial x_k}[m_k(\mathbf{x}, t)p] \\
&\quad + \sum_{j=1}^{2n}\sum_{k=1}^{2n} \frac{\partial^2}{\partial x_j \partial x_k}\left[\frac{b_{jk}(\mathbf{x}, t)}{2!}p\right] \\
&= \mathcal{L}_{FPK}[p(\mathbf{x}, t|\mathbf{x}_0, t_0)],
\end{aligned} \tag{8.10}
$$

subject to the initial condition

$$
p(\mathbf{x}, t_0|\mathbf{x}_0, t_0) = p(\mathbf{x}, t_0), \tag{8.11}
$$

where $\mathcal{L}_{FPK}[\cdot]$ stands for the differential operator of the FPK equation, $p(\mathbf{x}, t_0)$ is a given probability density function, and $b_{jk} = \sum_{l=1}^{m} \sigma_{kl}\sigma_{jl}$. $m_k(\mathbf{x}, t)$ and $b_{jk}(\mathbf{x}, t)$ are known as the drift and diffusion terms. Very often, the initial given is given as a delta function $p(\mathbf{x}, t_0) = \delta(\mathbf{x} - \mathbf{x}_0)$, i.e. the system starts at a point $\mathbf{x}_0$ with probability one.

To fully determine the probability density function $p(\mathbf{x}, t|\mathbf{x}_0, t_0)$ from the FPK equation, we need to specify an initial condition, and proper boundary conditions in the state space. In this work, we shall not consider boundary conditions. Hence, $\mathbf{x} \in \mathbb{R}^{2n}$.

We consider a trial solution for the FPK equation in the form of a RBFNN with Gaussian activation functions. For convenience, we denote the solution of the FPK equation $p(\mathbf{x}, t|\mathbf{x}_0, t_0)$ as $p(\mathbf{x}, t)$ subject to the initial condition $p(\mathbf{x}, t_0)$.

Let $\bar{p}(\mathbf{x}, \mathbf{w}(k))$ denote the solution $p(\mathbf{x}, t)$ at time step k where $\mathbf{w}(k)$ are a set of time-varying undetermined parameters. The solution $\bar{p}(\mathbf{x}, \mathbf{w}(k))$ consists of the RBFNN with Gaussian activation functions given by,

$$\bar{p}(\mathbf{x}, \mathbf{w}(k)) = \sum_{j=1}^{N_G} w_j(k) g(\mathbf{x}, \boldsymbol{\mu}_j, \boldsymbol{\Sigma}_j), \tag{8.12}$$

where N_G is the number of Gaussian functions, $\mathbf{x} \in \mathbb{R}^{2n}$ is the position vector. The model parameters w_j form a vector $\mathbf{w}(k) = [w_1(k), w_2(k), \ldots, w_{N_G}(k)]^T$. The time dependence of the assumed solution $\bar{p}$ is through that of the coefficients $\mathbf{w}(k)$. The index k indicates the time step such that $t = t_0 + k\tau$, $k = 0, 1, 2, \ldots$.

$g(\mathbf{x}, \boldsymbol{\mu}_j, \boldsymbol{\Sigma}_j)$ is a neuron in the form of a Gaussian function with the mean $\boldsymbol{\mu}_j$ and covariance matrix $\boldsymbol{\Sigma}_j$.

$$\begin{aligned} g(\mathbf{x}, \boldsymbol{\mu}_j, \boldsymbol{\Sigma}_j) = & \frac{1}{\sqrt{(2\pi)^{2n} \det(\boldsymbol{\Sigma}_j)}} \\ & \times \exp\left[-\frac{1}{2} (\mathbf{x} - \boldsymbol{\mu}_j)^T \boldsymbol{\Sigma}_j^{-1} (\mathbf{x} - \boldsymbol{\mu}_j) \right]. \end{aligned} \tag{8.13}$$

We should point out that the *radial* basis Gaussian function in the literature refers to the special case when $\boldsymbol{\Sigma}_j = \sigma_j^2 \mathbf{I}$, $\det(\boldsymbol{\Sigma}_j) = (\sigma_j^2)^{2n}$, $\boldsymbol{\Sigma}_j^{-1} = \sigma_j^{-2} \mathbf{I}$ and

$$\begin{aligned} g(\mathbf{x}, \boldsymbol{\mu}_j, \boldsymbol{\Sigma}_j) = & \frac{1}{\sqrt{(2\pi)^{2n} (\sigma_j^2)^{2n}}} \exp\left[-\frac{1}{2\sigma_j^2} ||\mathbf{x} - \boldsymbol{\mu}_j||^2 \right] \\ = & \prod_{i=1}^{2n} \frac{1}{\sqrt{2\pi}\sigma_j} \exp\left[-\frac{1}{2\sigma_j^2} (x_i - \mu_{j,i})^2 \right], \end{aligned} \tag{8.14}$$

where $\mathbf{I}$ is an identity matrix, and $|| \cdot ||$ denote the norm of a vector, and $\mu_{j,i}$ is the ith component of the mean vector $\boldsymbol{\mu}_j$. The radial form of the Gaussian functions makes computations a lot easier and is used in the numerical studied reported later.

It is noted that the transient probability density function depends on the initial condition. When the initial probability distribution is highly concentrated such as a delta function $\delta(\mathbf{x} - \mathbf{x}_0)$, we can use the short-time analytical approximate solution to compute the probability density over one or a few time steps τ to avoid the need for extremely small meshes.

The RBFNN PDF solution must satisfy the normalization condition,

$$\int_{\mathbb{R}^{2n}} \bar{p}(\mathbf{x}, \mathbf{w}(k))d\mathbf{x} = 1, \quad \text{for } k \geq 0. \tag{8.15}$$

Because of the proper definition of the Gaussian activation functions such that

$$\int_{\mathbb{R}^{2n}} g(\mathbf{x}, \boldsymbol{\mu}_j, \boldsymbol{\Sigma}_j)d\mathbf{x} = 1, \quad \text{for all } j, \tag{8.16}$$

the normalization condition for the RBFNN PDF solution becomes

$$\sum_{j=1}^{N_G} w_j(k) = 1, \quad \text{for } k \geq 0. \tag{8.17}$$

Recall the initial condition $p_0(\mathbf{x})$ of the FPK Eq. (8.11). We apply the solution given in Eq. (8.12) with a proper choice of the mean $\boldsymbol{\mu}_j$ and covariance matrix $\boldsymbol{\Sigma}_j$. We obtain the initial coefficients $\mathbf{w}(0)$. Let τ be a small time step for integrating the FPK equation forward in time, and $\mathbf{w}(k)$ be the coefficient vector at time $t = k\tau$, $k = 0, 1, 2, \ldots$.

We apply the finite difference to compute the time derivative

$$\frac{\partial \bar{p}}{\partial t} = \frac{1}{\tau}\left[\bar{p}(\mathbf{x}, \mathbf{w}(k+1)) - \bar{p}(\mathbf{x}, \mathbf{w}(k))\right] + \epsilon_t, \tag{8.18}$$

where ϵ_t is the truncation error of the first order finite difference approximation.

Substitute the solution in Eq. (8.12) and the time derivative in Eqs. (8.18)–(8.10). We define a local error of the equation as

$$\begin{aligned} e(\mathbf{x}, \mathbf{w}(k)) &= -\bar{p}(\mathbf{x}, \mathbf{w}(k)) + \bar{p}(\mathbf{x}, \mathbf{w}(k-1)) \\ &\quad + \mathcal{L}_{FPK}[\bar{p}(\mathbf{x}, \mathbf{w}(k))]\tau \\ &= \bar{p}(\mathbf{x}, \mathbf{w}(k-1)) + \sum_{j=1}^{N_G} s_j(\mathbf{x}) w_j(k), \end{aligned} \tag{8.19}$$

where $k = 1, 2, \ldots$ and

$$s_j(\mathbf{x}) = -g(\mathbf{x}, \boldsymbol{\mu}_j, \boldsymbol{\Sigma}_j) + \tau \mathcal{L}_{FPK}[g(\mathbf{x}, \boldsymbol{\mu}_j, \boldsymbol{\Sigma}_j)]. \tag{8.20}$$

The explicit expression for $s_j(\mathbf{x})$ after all the differentiations are carried out can be obtained symbolically in order to speed up the computation. Note that at step k, $\bar{p}(\mathbf{x}, \mathbf{w}(k-1))$ is fully determined. The truncation error ϵ_t can be thought of as being absorbed in the error $e(\mathbf{x}, \mathbf{w}(k))$.

We define an integrated squared error together with a Lagrange multiplier $\lambda(k)$ for the normalization condition in Eq. (8.17) as

$$J(\mathbf{w}(k), \lambda(k)) = \int\limits_{\mathbb{R}^{2n}} \frac{1}{2} e^2(\mathbf{x}, \mathbf{w}(k)) d\mathbf{x} + \lambda(k) \left(\sum_{j=1}^{N_G} w_j(k) - 1 \right). \tag{8.21}$$

We should point out that this step of optimization is important to correct the truncation error so that at every integration time step, the solution is obtained in the sense of the least squared error.

We select $\mathbf{w}(k)$ and $\lambda(k)$ to minimize J. Define a parameter vector as

$$\begin{aligned} \mathbf{c}(k) &= [w_1(k), w_2(k), \ldots, w_{N_G}(k), \lambda(k)]^T \\ &= [\mathbf{w}^T(k), \lambda(k)]^T. \end{aligned} \tag{8.22}$$

For convenience, we shall denote $J(\mathbf{w}(k), \lambda(k))$ as $J(\mathbf{c}(k))$. We shall derive necessary conditions for minimization of $J(\mathbf{c}(k))$.

8.8 Motivation for Gaussian Functions

It is well known that the generalized cell mapping (GCM) with the short-time Gaussian approximation (STGA) is a highly effective method for computing the probabilistic response of nonlinear stochastic systems [10, 11]. In the GCM-STGA method, the one-step transition probability matrix of nonlinear stochastic systems starting from each initial cell has a Gaussian distribution over a short time. The probabilistic response of the stochastic system is fully determined by the one-step transition probability matrix from all the initial cells in a given domain. The GCM-STGA method indicates that at every mapping step, the probabilistic distribution of the response is a linear combination of many Gaussian probability distributions. This is consistent with the RBFNN solution given in Eq. (8.12).

8.9 Sampling Method for Integration

Note that the computation of $J(\mathbf{c}(k))$ involving integrations in high dimensional space can be intensive. Instead of integration, we can randomly sample a large number of points $\mathbf{x}_i \in D$ where $D \subset \mathbb{R}^{2n}$ is a large enough domain in order to calculate an approximate value of the integrated squared error as

$$J_s(\mathbf{c}(k)) = \sum_{i=1}^{N_s} \frac{1}{2} e^2(\mathbf{x}_i, \mathbf{w}(k)) + \lambda(k) \left(\sum_{j=1}^{N_G} w_j(k) - 1 \right), \quad \mathbf{x}_i \in D \tag{8.23}$$

where N_s is the number of sampled points. The sampled performance index $J_s(\mathbf{c})$ can be written in the following form.

$$\begin{aligned} J_s(\mathbf{c}(k)) &= \sum_{i=1}^{N_s} \frac{1}{2} \left(\sum_{j=1}^{N_G} \sum_{l=1}^{N_G} s_j(\mathbf{x}_i) s_l(\mathbf{x}_i) w_j(k) w_l(k) \right) \\ &+ \sum_{i=1}^{N_s} \sum_{j=1}^{N_G} s_j(\mathbf{x}_i) \bar{p}(\mathbf{x}_i, \mathbf{w}(k-1)) w_j(k) \\ &+ \frac{1}{2} \sum_{i=1}^{N_s} \bar{p}^2(\mathbf{x}_i, \mathbf{w}(k-1)) + \lambda(k) \left(\sum_{j=1}^{N_G} w_j(k) - 1 \right) \\ &\equiv \mathbf{c}^T \mathbf{A} \mathbf{c}(k) + \mathbf{c}^T [\mathbf{b}_p(k) - \mathbf{b}_\lambda] + d(k-1), \end{aligned} \tag{8.24}$$

where the matrix $\mathbf{A} \in \mathbb{R}^{(N_G+1)\times(N_G+1)}$ is defined with the help of a data matrix $\mathbf{S} \in \mathbb{R}^{N_G \times N_s}$. The vectors $\mathbf{b}_\lambda$ and $\mathbf{b}_p \in \mathbb{R}^{(N_G+1)\times 1}$ together with a scalar $d(k-1)$ are also defined below.

$$\mathbf{S} = [S_{jl}] = \left[s_j(\mathbf{x}_l) \right], \tag{8.25}$$

$$\mathbf{G} = [G_{jl}] = \left[g_j(\mathbf{x}_l) \right], \tag{8.26}$$

$$\mathbf{b}_\lambda = [0, 0, \ldots, 0, 1]^T, \tag{8.27}$$

$$\mathbf{A} = \begin{bmatrix} & 1 \\ \frac{1}{2}\mathbf{S}\mathbf{S}^T & \vdots \\ & 1 \\ 0 \cdots 0 & 0 \end{bmatrix}, \tag{8.28}$$

$$\mathbf{B} = \begin{bmatrix} & 0 \\ \mathbf{S}\mathbf{G}^T & \vdots \\ & 0 \\ 0 \cdots 0 & 0 \end{bmatrix}, \tag{8.29}$$

and

$$
\begin{aligned}
b_{pj}(k) &= \sum_{i=1}^{N_s} s_j(\mathbf{x}_i)\bar{p}(\mathbf{x}_i, \mathbf{w}(k-1)) \\
&= \sum_{i=1}^{N_s} s_j(\mathbf{x}_i) \sum_{l=1}^{N_G} w_l(k-1) G(\mathbf{x}_i, \boldsymbol{\mu}_l, \boldsymbol{\Sigma}_l),
\end{aligned} \tag{8.30}
$$

$$
\mathbf{b}_p(k) = [b_{p1}, b_{p2}, \ldots, b_{pN_G}, 0]^T = \mathbf{Bc}(k-1), \tag{8.31}
$$

$$
d(k-1) = \sum_{i=1}^{N_s} \bar{p}^2(\mathbf{x}_i, \mathbf{w}(k-1)). \tag{8.32}
$$

The necessary condition for minimizing J_s with respect to the vector $\mathbf{c}$ reads

$$
\frac{\partial J}{\partial \mathbf{c}(k)} = (\mathbf{A} + \mathbf{A}^T)\mathbf{c}(k) + \mathbf{b}_p(k) - \mathbf{b}_\lambda = 0. \tag{8.33}
$$

The optimal coefficient vector is given by

$$
\mathbf{c}_{opt}(k) = (\mathbf{A} + \mathbf{A}^T)^{-1}[\mathbf{b}_\lambda - \mathbf{Bc}(k-1)], \tag{8.34}
$$

when the inverse of the matrix $(\mathbf{A} + \mathbf{A}^T)$ exists. The existence of the inversion heavily depends on the so-called data matrix $\mathbf{S}$. As is well-known in the regression research [12], the data matrix also determines the overall performance of the RBFNN PDF solution in Eq. (8.12).

About Sampling

The purpose of adopting the sampling method to compute integration in high dimensional state space is not simply to avoid the curse of dimensionality of numerical integration in high dimensional state space. The sampling process can mimic the situation when the system is not completely known and sampled data from field measurements are available. Hence, the RBFNN solution can be data-driven. In fact, the best results can be obtained when both the data and knowledge about the physical model of the system are used to compute the PDF solution.

8.10 Response Moments

With the probability density function on hand, many response statistics can be computed including the moments. The first and second order moments of the system response at the kth time step can be computed as follows.

$$\begin{aligned}
\mathrm{E}[\mathbf{X}](k) &= \int\limits_{\mathbb{R}^{2n}} \mathbf{x}\bar{p}(\mathbf{x}, \mathbf{w}(k))d\mathbf{x} \\
&= \sum_{j=1}^{N_G} w_j(k) \int\limits_{\mathbb{R}^{2n}} \mathbf{x}g(\mathbf{x}, \boldsymbol{\mu}_j, \boldsymbol{\Sigma}_j)d\mathbf{x} \\
&= \sum_{j=1}^{N_G} w_j(k)\boldsymbol{\mu}_j.
\end{aligned} \tag{8.35}$$

The expectation with respect to the jth Gaussian function has the following property.

$$\begin{aligned}
\mathrm{E}[\mathbf{X}^T\mathbf{X}] = &\mathrm{E}[(\mathbf{X} - \boldsymbol{\mu}_j)^T(\mathbf{X} - \boldsymbol{\mu}_j)] \\
&+ \mathrm{E}[(\mathbf{X} - \boldsymbol{\mu}_j)^T]\boldsymbol{\mu}_j \\
&+ \boldsymbol{\mu}_j\mathrm{E}[(\mathbf{X} - \boldsymbol{\mu}_j)^T] + \boldsymbol{\mu}_j^T\boldsymbol{\mu}_j.
\end{aligned} \tag{8.36}$$

By definition, we have

$$\begin{aligned}
&\mathrm{E}[(\mathbf{X} - \boldsymbol{\mu}_j)^T(\mathbf{X} - \boldsymbol{\mu}_j)] = \boldsymbol{\Sigma}_j, \\
&\mathrm{E}[(\mathbf{X} - \boldsymbol{\mu}_j)] = 0.
\end{aligned} \tag{8.37}$$

Hence, we have the correlation matrix, i.e. the second order moments, as

$$\begin{aligned}
\mathrm{E}[\mathbf{X}^T\mathbf{X}](k) &= \int\limits_{\mathbb{R}^{2n}} \mathbf{x}^T\mathbf{x}\bar{p}(\mathbf{x}, \mathbf{w}(k))d\mathbf{x} \\
&= \sum_{j=1}^{N_G} w_j(k)[\boldsymbol{\Sigma}_j + \boldsymbol{\mu}_j^T\boldsymbol{\mu}_j].
\end{aligned} \tag{8.38}$$

Once the first and second moments are obtained, we can compute the response moments of any order. Let us use a typical moment $\mathrm{E}[X_i^p X_k^q]$ to illustrate the computational process by using the radial basis Gaussian functions in Eq. (8.14). Because of the special radial form of the Gaussian functions, it implies that for the jth Gaussian distribution $g(\mathbf{x}, \boldsymbol{\mu}_j, \boldsymbol{\Sigma}_j)$, the random variables X_i and X_k can be viewed as being independent for $i \neq k$. Hence, the expectation with respect to this Gaussian distribution can be written as

$$\mathrm{E}[X_i^p X_k^q] = \mathrm{E}[X_i^p] \cdot \mathrm{E}[X_k^q]. \tag{8.39}$$

That is, we only need to focus on the computation of the term $\mathrm{E}[X_i^p]$. To compute this term, we convert it to the central moment as

$$\mathrm{E}[(X_i - \mu_{j,i} + \mu_{j,i})^p] = \sum_{l=0}^{p} C_l^p \mathrm{E}[(X_i - \mu_{j,i})^{p-l}]\mu_{j,i}^l, \tag{8.40}$$

where

$$C_l^p = \frac{p!}{(p-l)!l!}. \tag{8.41}$$

The central moments determined by the jth Gaussian function $g(\mathbf{x}, \boldsymbol{\mu}_j, \boldsymbol{\Sigma}_j)$ are given by

$$\mathrm{E}[(X_i - \mu_{j,i})^p] = \begin{cases} 0, & \text{if } p \text{ is odd,} \\ \sigma_j^p (p-1)!, & \text{if } p \text{ is even.} \end{cases} \tag{8.42}$$

With the help of this result, arbitrary cross moments such as $\mathrm{E}[X_i^p X_k^q]$ can be expressed in terms of lower order moments with respect to each Gaussian function.

This demonstrates another advantage of choosing the true probability density as the basis function. The response moments of any order can be computed without numerical integration in the space $\mathbb{R}^{2n}$.

8.11 Examples of Application

8.11.1 First Order Nonlinear System

Consider a first order nonlinear system

$$\frac{dX}{dt} = aX - bX^3 + W_1(t), \tag{8.43}$$

where $b > 0$, and $W_1(t)$ denotes the Gaussian white noise excitation with intensity $2D_1$. The drift and diffusion coefficients of the FPK equation are given by,

$$m_1(x,t) = ax - bx^3, \quad b_{22} = 2D_1. \tag{8.44}$$

We choose the computational domain in the state space to be $D = [-5, 5]$ and the parameters $a = 1$, $b = 2$ and $D_1 = 3$. The number of Gaussian functions is $N_G = 101$. The means of all the Gaussian functions are at the grid points, while the standard deviations of all the Gaussian functions are taken to be the same constant $\sigma_j = h$ where h is the grid size $h = 10/101$. The number of sampling point $N_s = 2N_G$, which was found to be sufficient. We should point out that when $a > 0$, the steady-state PDF function of the system is bi-modal because $x = 0$ is an unstable equilibrium.

Consider an initial condition as $x_0 = 0$ at $t = 0$ with probability one. We choose a small time step for integration as $\tau = 0.01$, and construct a short-time Gaus-

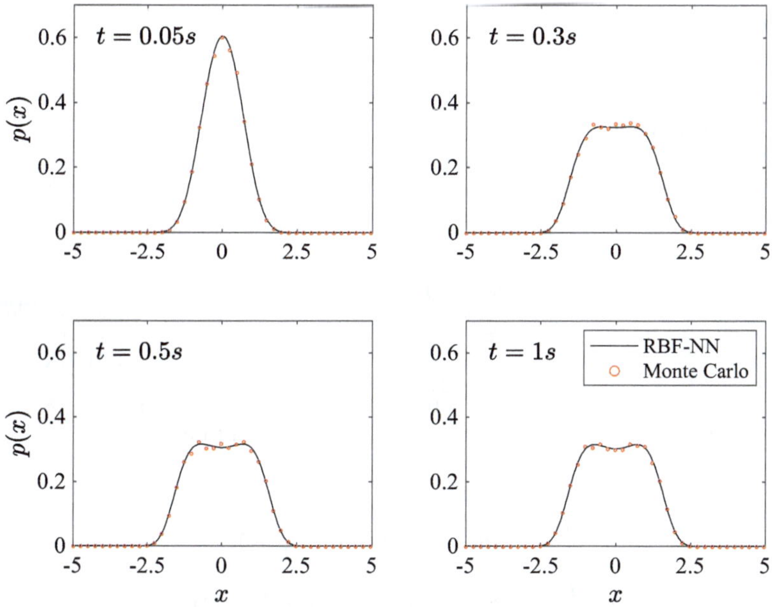

Fig. 8.7 The evolution of the PDF of system (8.43) obtained by the proposed method and Monte Carlo simulation

sian approximation of the PDF with the mean $x_0 + m(x_0, 0)\tau = 0$ and variance $2b(x_0, 0)\tau = 0.12$.

Figure 8.7 presents evolutions of the response PDF and a comparison of the RBFNN PDF solution with Monte Carlo simulations. In this example, we generate 10^5 sample time histories in simulations in order to compute all the response statistics reported herein. Simulations take 103 s. As a comparison, the RBFNN method takes 2.83 s to compute the same set of solutions. The overall rms error of the RBFNN PDF solution compared to Monte Carlo simulation is 0.0141. The agreement between the two results is indeed excellent.

Let $m_p = \text{E}[X^p]$ denote the pth order moment of the response. We apply Eqs. (8.35) and (8.38) to compute the first and second order moments of the response. Furthermore, we shall compare the moments obtained by the RBFNN PDF solution with those by Monte Carlo simulations and Gaussian closure approximation.

The Gaussian closure makes use of the following relationships of higher order moments to the lower order ones.

$$
\begin{aligned}
m_3 &= 3m_2 m_1 - 2m_1^3, \\
m_4 &= 3m_2^2 - 2m_1^4.
\end{aligned}
\tag{8.45}
$$

The equations for the first and second order moments of system (8.43) obtained with the Gaussian closure are given by

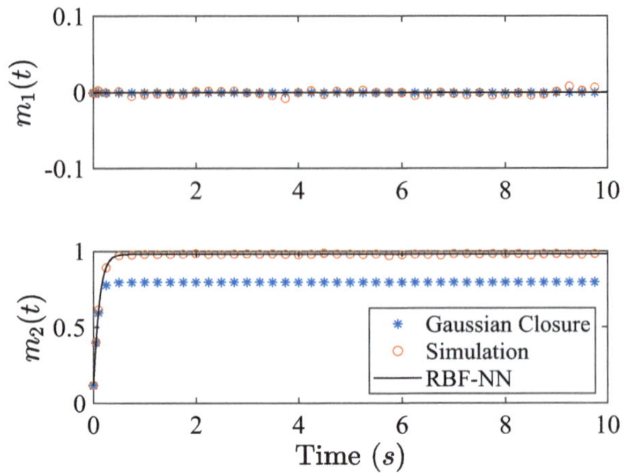

Fig. 8.8 The first and second order moments of the system response by the proposed method, Gaussian closure and Monte Carlo simulations. $a = 1$

$$\frac{dm_1}{dt} = am_1 - b(3m_2m_1 - 2m_1^3),$$
$$\frac{dm_2}{dt} = 2am_2 - 6bm_2^2 + 4bm_1^4 + 2D_1. \tag{8.46}$$

We first consider a case when $a = 1 > 0$ and the steady-state PDF function of the system is bi-modal. The bi-modal distribution is certainly far away from being Gaussian. Figure 8.8 shows the time histories of the first and second order moments of the system response by the RBFNN solution, the Gaussian closure method and Monte Carlo simulations. It is seen from the figure that the response moments obtained with the PDF of the RBFNN solution is highly accurate as compared to Monte Carlo simulations, while the Gaussian closure performs poorly because the underlining process is far away from being Gaussian.

The Gaussian closure method is only good for weakly nonlinear system when the system response distribution is nearly Gaussian. When the parameter $a < 0$, the response distribution of system (8.43) is nearly Gaussian with a single peak at $x = 0$. Let $a = -1$, $b = 0.1$ and $D_1 = 3$. Figure 8.9 shows the first and second order response moments of the system by the the RBFNN solution, the Gaussian closure method and Monte Carlo simulations. The performance of the Gaussian closure method improves significantly because the response of the system is nearly Gaussian.

Figure 8.10 shows the steady-state second order moments of the system response as a function of the parameter b. It should be noted that when b is large, even when $a > 0$, the two-peaks of the probability distribution move closer to each other, resulting a distribution which looks more like a Gaussian function. Hence, as b increases, the error of the Gaussian closure solution decreases. The RBFNN solution is consistently accurate as compared to Monte Carlo simulations for all the parameters we considered.

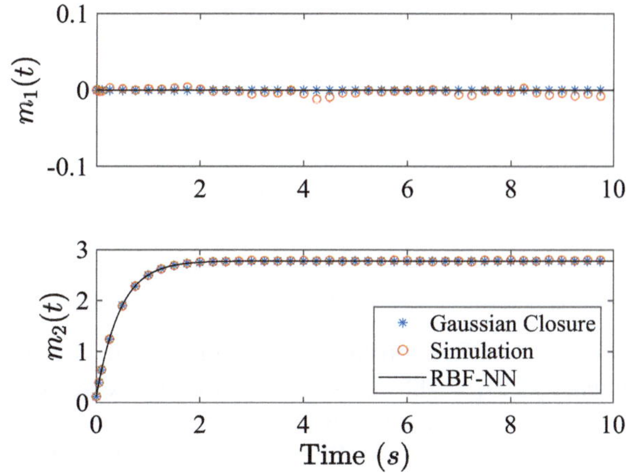

Fig. 8.9 The first and second order moments of the system response by the proposed method, Gaussian closure and Monte Carlo simulations. $a = -1$

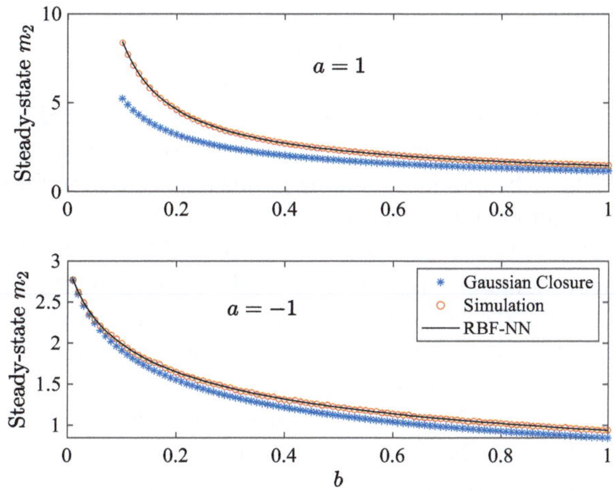

Fig. 8.10 The steady-state second order moments of the system response as a function of the parameter b by the proposed method, Gaussian closure and Monte Carlo simulations

In Fig. 8.10 for $a = 1$, the average error in m_2 over the range of parameter b of the RBFNN solution is 0.2580%, while the same error of the Gaussian closure solution is 24.7659%, which is about 96 times bigger than that of the RBFNN solution. For $a = -1$, the average error in m_2 over the range of parameter b of the RBFNN solution is 0.2948%, while the same error of the Gaussian closure solution is 7.4546%, which is about 25 times bigger than that of the RBFNN solution. In this case, the response is close to be Gaussian.

8.11.2 Van der Pol Oscillator

Consider a strongly nonlinear Van der Pol oscillator under external Gaussian white noise excitation.

$$
\begin{aligned}
\frac{dX}{dt} &= Y, \\
\frac{dY}{dt} &= -\beta(X^2 - 1)Y - X + W_1(t),
\end{aligned}
\tag{8.47}
$$

where $\beta = 1.5$ is constant. $W_1(t)$ denotes the Gaussian white noise excitation with intensity $2D_1 = 0.5$. This system has long been a challenging example in the study of stochastic dynamic systems. There has been no report in the literature on the transient response of the PDF. Here, we apply the RBFNN solution method to obtain the transient probability distribution of the system response and the time-varying moments.

Let us take the domain of interest in the state space as $D_G = [-3, 3] \times [-5.5, 5.5]$ and partition it into a $N_G \times N_G$ grid where $N_G = 100$. At each grid, there is a Gaussian function. Every Gaussian function is assumed to have the same standard deviation $\sigma_j = h$ where h is the grid size. We have found from computational exercises that it is beneficial to sample points for computing the data matrix in a slightly larger domain than D_G. In this example, we take $D_s = [-4, 4] \times [-7, 7]$ and uniformly sample $N_s \times N_s$ points in D_s where $N_s = 200$. We use the short-time Gaussian approximation of the PDF to deal with the initial condition with probability one at $(x, y) = (0, 0)$ when $t = 0$. When $t = \tau$, the response distribution is approximately Gaussian with the mean $[0, 0]$ and variance $[0.05, 0.05]$, according to the short-time Gaussian approximation.

Figure 8.11 compares the steady-state PDFs of the Van der Pol oscillator obtained by the RBFNN solution method and Monte Carlo simulations. 10^6 samples are generated in simulations, which takes 1555.3 s. As a comparison, the RBFNN method takes 2226.6 s when $N_G = 100$ and $N_s = 200$, and 540 s when $N_G = 60$ and $N_s = 120$. The overall rms error of the RBFNN PDF solution compared to Monte Carlo simulation is 0.0196. We shall further examine the accuracy of the solution by considering the solutions of the response moments.

Let $m_{pq} = \mathrm{E}[X^p Y^q]$. The equations governing the moment evolution of the Van der Pol oscillator after applying Gaussian closure are obtained as,

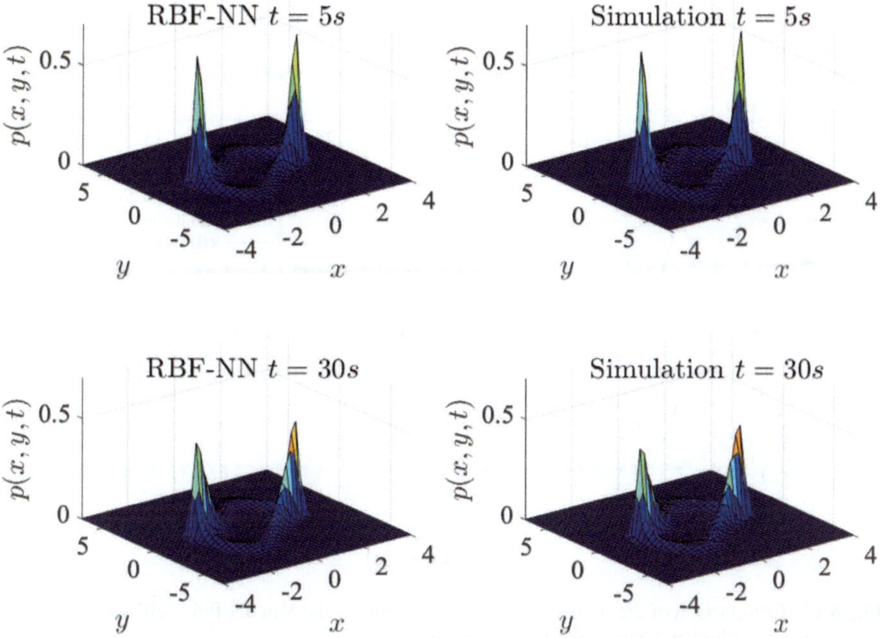

Fig. 8.11 Comparison of the PDF solution of the Van der Pol oscillator obtained by the proposed method and Monte Carlo simulations

$$\begin{aligned}
\frac{dm_{10}}{dt} &= m_{01}, \\
\frac{dm_{01}}{dt} &= -\beta(m_{20} - 1)m_{01} - m_{10}, \\
\frac{dm_{20}}{dt} &= 2m_{11}, \\
\frac{dm_{11}}{dt} &= -\beta(3m_{20}m_{11} - 2m_{10}^3 m_{01}) + \beta m_{11} - m_{20} + m_{02}, \\
\frac{dm_{02}}{dt} &= -2\beta m_{20}m_{02} + 2\beta m_{02} - 2m_{11} + 2D_1.
\end{aligned} \tag{8.48}$$

We shall not show the first order response moments of the system because the solutions obtained by all the methods under consideration are nearly zero as they should be. Figure 8.12 shows the second order response moments of the system by the RBFNN solution, the Gaussian closure method and Monte Carlo simulations. It is seen from the figure that the RBFNN solution is consistently accurate as compared to Monte Carlo simulations, while m_{20} and m_{02} obtained by the Gaussian closure method have significant steady-state error in addition to poor transient performance. The reason for poor performance of the Gaussian closure is explained by the PDF

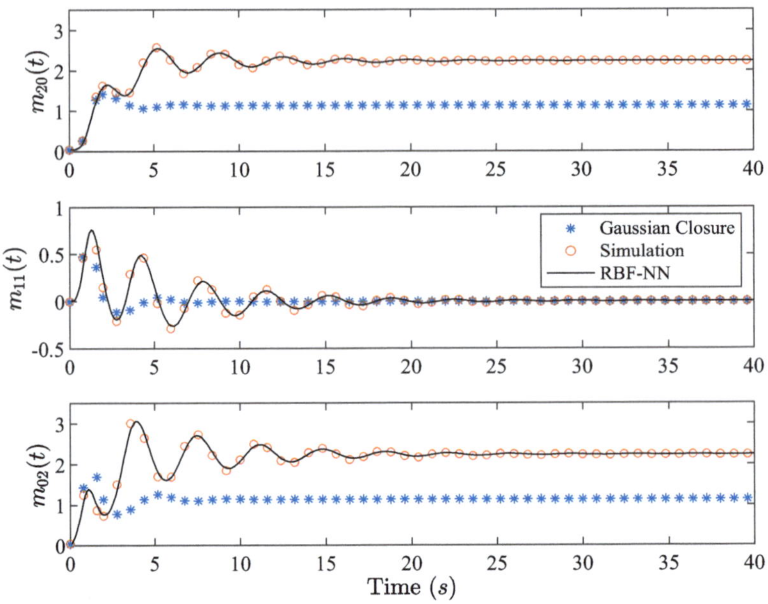

Fig. 8.12 Comparison of the transient response moments of the Van der Pol oscillator obtained by the proposed method and Monte Carlo simulations

shapes in Fig. 8.11. For $t > 0$ away from the concentrated initial condition, the PDF of the Van der Pol oscillator is far away from being Gaussian.

However, at any time $t > 0$, the geometrically complex PDF of the Van der Pol oscillator can always be expressed as a linear combination of radial basis Gaussian functions given in Eq. (8.12) with a proper choice of the means and standard deviations. Furthermore, when N_G is sufficiently large, the error of the RBFNN solution converges to zero [8].

Let us now discuss the solution errors of response moments in reference to the simulation results. The steady-state error averaged over the last 100 data points in m_{20} of the RBFNN solution is 0.00079333, which represents a very small percentage of the response moment. The same error of the Gaussian closure solution is 1.0908. This is over 40% of the response moment m_{20} according to the simulation results. The steady-state error averaged over the last 100 data points in m_{02} of the RBFNN solution is 0.0018222, which is also a very small percentage of the response moment. The same error of the Gaussian closure solution is still 1.0908, and is over 40% of the response moment m_{02}. The steady-state error averaged over the last 100 data points in m_{11} is 0.00072965 of the RBFNN solution, and 0.00080836 of the Gaussian closure solution. The exact steady-state value of m_{11} is known to be zero.

8.12 First-Passage Problem

Recall the backward Kolmogorov Eq. (6.21).

$$
\begin{aligned}
-\frac{\partial}{\partial t_0} p_{\mathbf{X}}(\mathbf{x}, t|\mathbf{x}_0, t_0) &= \sum_{j=1}^{2n} A_j(\mathbf{x}_0, t_0) \frac{\partial p_{\mathbf{X}}}{\partial x_{0j}} \\
&+ \sum_{j=1}^{2n} \sum_{k=1}^{2n} \frac{B_{jk}(\mathbf{x}_0, t_0)}{2!} \frac{\partial^2 p_{\mathbf{X}}}{\partial x_{0j} x_{0k}} \\
&\equiv \mathcal{L}_{BK}[p(\mathbf{x}, t|\mathbf{x}_0, t_0)],
\end{aligned} \tag{8.49}
$$

where $\mathcal{L}_{BK}[\cdot]$ denotes the differential operator of the backward Kolmogorov equation. Other terms are defined in Sect. 6.2.

When $\mathbf{X}(t)$ is a stationary process such that $p_{\mathbf{X}}(\mathbf{x}, t|\mathbf{x}_0, t_0) = p_{\mathbf{X}}(\mathbf{x}, t - t_0|\mathbf{x}_0)$, the reliability satisfies the following equation.

$$
\frac{\partial}{\partial \tau} R_{\mathcal{S}}(\tau, \mathbf{x}_0) = \mathcal{L}_{BK}[R_{\mathcal{S}}(\tau, \mathbf{x}_0)], \tag{8.50}
$$

$$
R_{\mathcal{S}}(0, \mathbf{x}_0) = 1, \qquad \mathbf{x}_0 \in \mathcal{S}, \tag{8.51}
$$

$$
R_{\mathcal{S}}(\tau, \mathbf{x}_0) = 0, \qquad \mathbf{x}_0 \in \partial\mathcal{S}, \tag{8.52}
$$

where $\tau = t - t_0$ and $\mathcal{S} \subset \mathbb{R}^{2n}$ is the safe domain. The reliability is therefore a solution of the partial differential equation subject to an initial and boundary condition.

8.12.1 RBFNN Solution

Let $\Delta\tau$ be a small integration time step. At the kth step, we denote the reliability $R_{\mathcal{S}}(k\Delta\tau, \mathbf{x}_0)$ as $R(\mathbf{x}_0, \mathbf{w}(k))$. We express $R(\mathbf{x}_0, \mathbf{w}(k))$ in terms of radial basis function neural networks (RBFNN) with Gaussian activation functions.

$$
R(\mathbf{x}_0, \mathbf{w}(k)) = \sum_{j=1}^{N_G} w_j(k) g(\mathbf{x}_0, \boldsymbol{\mu}_j, \boldsymbol{\Sigma}_j), \tag{8.53}
$$

where N_G is the number of Gaussian functions, $\mathbf{x}_0 \in \mathcal{S}$. The model parameters w_j form a vector $\mathbf{w}(k) = [w_1(k), w_2(k), \ldots, w_{N_G}(k)]^T$. The time dependence of the solution is through that of the coefficients $\mathbf{w}(k)$. $g(\mathbf{x}_0, \boldsymbol{\mu}_j, \boldsymbol{\Sigma}_j)$ is a neuron in the form of a Gaussian function with the mean $\boldsymbol{\mu}_j$ and covariance matrix $\boldsymbol{\Sigma}_j$.

$$g(\mathbf{x}_0, \boldsymbol{\mu}_j, \boldsymbol{\Sigma}_j) = \frac{1}{\sqrt{(2\pi)^{2n}(\sigma_j^2)^{2n}}} \exp\left[-\frac{1}{2\sigma_j^2}||\mathbf{x}_0 - \boldsymbol{\mu}_j||^2\right]$$
$$= \prod_{i=1}^{2n} \frac{1}{\sqrt{2\pi}\sigma_j} \exp\left[-\frac{1}{2\sigma_j^2}(x_{i0} - \mu_{j,i})^2\right], \tag{8.54}$$

where $\boldsymbol{\Sigma}_j = \sigma_j^2\mathbf{I}$, $\det(\boldsymbol{\Sigma}_j) = (\sigma_j^2)^{2n}$, $\boldsymbol{\Sigma}_j^{-1} = \sigma_j^{-2}\mathbf{I}$, $\mathbf{I}$ is an identity matrix, $||\cdot||$ denotes the norm of a vector, and $\mu_{j,i}$ is the ith component of the mean vector $\boldsymbol{\mu}_j$. The radial form of the Gaussian functions makes computations much simpler.

We apply the finite difference to compute the time derivative of the reliability function

$$\frac{\partial R}{\partial \tau} = \frac{1}{\Delta\tau}[R(\mathbf{x}_0, \mathbf{w}(k)) - R(\mathbf{x}_0, \mathbf{w}(k-1))] + \epsilon_\tau, \tag{8.55}$$

where ϵ_τ is the truncation error of the first order finite difference approximation.

Substitute the solution in Eq. (8.53) and the time derivative in Eqs. (8.55)–(8.50). We define a local error of the equation as

$$e(\mathbf{x}_0, \mathbf{w}(k)) = -R(\mathbf{x}_0, \mathbf{w}(k)) + R(\mathbf{x}_0, \mathbf{w}(k-1))$$
$$+ \mathcal{L}_{BK}[R(\mathbf{x}_0, \mathbf{w}(k))]\Delta\tau$$
$$= R(\mathbf{x}_0, \mathbf{w}(k-1)) + \sum_{j=1}^{N_G} s_j(\mathbf{x}_0) w_j(k), \tag{8.56}$$

where

$$s_j(\mathbf{x}_0) = -g(\mathbf{x}_0, \boldsymbol{\mu}_j, \boldsymbol{\Sigma}_j) + \Delta\tau\mathcal{L}_{BK}[g(\mathbf{x}_0, \boldsymbol{\mu}_j, \boldsymbol{\Sigma}_j)]. \tag{8.57}$$

Note that at step k, $R(\mathbf{x}_0, \mathbf{w}(k-1))$ is fully determined. The truncation error ϵ_τ is absorbed in the error $e(\mathbf{x}_0, \mathbf{w}(k))$.

We define an integrated squared error in the domain $\mathcal{S}$ as

$$J_{\mathcal{S}}(\mathbf{w}(k)) = \int_{\mathcal{S}} \frac{1}{2} e^2(\mathbf{x}_0, \mathbf{w}(k)) d\mathbf{x}_0, \tag{8.58}$$

and the integrated squared error on the boundary condition as

$$J_{\partial\mathcal{S}}(\mathbf{w}(k)) = \int_{\partial\mathcal{S}} \frac{1}{2} R^2(\mathbf{x}_0, \mathbf{w}(k)) d\mathbf{x}_0. \tag{8.59}$$

Applying the method of Lagrange's multiplier, we define the total error as

$$J(\mathbf{w}(k), \lambda(k)) = J_{\mathcal{S}}(\mathbf{w}(k)) + \lambda(k) J_{\partial\mathcal{S}}(\mathbf{w}(k)). \tag{8.60}$$

The computation of $J(\mathbf{w}(k), \lambda(k))$ involves integration in high dimensional space and can be intensive. Instead of integration, we can uniformly sample a large number of points $\mathbf{x}_{0i} \in \mathcal{S}$ and $\mathbf{y}_{0j} \in \partial\mathcal{S}$. Hence, we obtain a sampled value J_s of J as

$$J_s(\mathbf{w}(k), \lambda(k)) = \sum_{i=1}^{N_s} \frac{1}{2} e^2(\mathbf{x}_{0i}, \mathbf{w}(k)) + \lambda(k) \sum_{j=1}^{M_s} \frac{1}{2} R^2(\mathbf{y}_{0j}, \mathbf{w}(k)), \tag{8.61}$$

where $\mathbf{x}_{0i} \in \mathcal{S}$, $\mathbf{y}_{0j} \in \partial\mathcal{S}$, N_s is the number of sampled points in the domain $\mathcal{S}$ and M_s is the number of sampled points on the boundary $\partial\mathcal{S}$. The sampled performance index reads,

$$\begin{aligned}\sum_{i=1}^{N_s} \frac{1}{2} e^2(\mathbf{x}_{0i}, \mathbf{w}(k)) &= \sum_{i=1}^{N_s} \frac{1}{2} \left(\sum_{j=1}^{N_G} \sum_{l=1}^{N_G} s_j(\mathbf{x}_{0i}) s_l(\mathbf{x}_i) w_j(k) w_l(k) \right) \\ &+ \sum_{i=1}^{N_s} \sum_{j=1}^{N_G} s_j(\mathbf{x}_{0i}) R(\mathbf{x}_{0i}, \mathbf{w}(k-1)) w_j(k) \\ &+ \frac{1}{2} \sum_{i=1}^{N_s} R^2(\mathbf{x}_{0i}, \mathbf{w}(k-1)) \\ &\equiv \frac{1}{2} \mathbf{w}^T \mathbf{A} \mathbf{w}(k) + \mathbf{w}^T \mathbf{b}(k) + d(k-1),\end{aligned} \tag{8.62}$$

and

$$\begin{aligned}&\sum_{j=1}^{M_s} \frac{1}{2} R^2(\mathbf{y}_{0j}, \mathbf{w}(k)) \\ &= \sum_{j=1}^{M_s} \frac{1}{2} \left(\sum_{i=1}^{N_G} \sum_{l=1}^{N_G} g_i(\mathbf{y}_{0j}, \boldsymbol{\mu}_i, \boldsymbol{\Sigma}_i) g_l(\mathbf{y}_{0j}, \boldsymbol{\mu}_l, \boldsymbol{\Sigma}_l) w_i(k) w_l(k) \right) \\ &\equiv \frac{1}{2} \mathbf{w}^T \mathbf{R}_b \mathbf{w}(k),\end{aligned} \tag{8.63}$$

where the matrix $\mathbf{A} \in \mathbb{R}^{N_G \times N_G}$, $\mathbf{B} \in \mathbb{R}^{N_G \times N_G}$ are defined with the help of data matrix $\mathbf{S} \in \mathbb{R}^{N_G \times N_s}$, $\mathbf{G} \in \mathbb{R}^{N_G \times N_s}$, the vector $\mathbf{b} \in \mathbb{R}^{N_G \times 1}$, $\mathbf{R}_b \in \mathbb{R}^{N_G \times N_G}$ and $d(k-1)$ is a scalar. These are all defined below.

$$\mathbf{G} = [G_{li}] = \left[g(\mathbf{x}_{0i}, \boldsymbol{\mu}_l, \boldsymbol{\Sigma}_l)\right], \mathbf{x}_{0i} \in \mathcal{S} \tag{8.64}$$

$$\mathbf{G}_b = [G_{lj}] = \left[g(\mathbf{y}_{0j}, \boldsymbol{\mu}_l, \boldsymbol{\Sigma}_l)\right], \mathbf{y}_{0j} \in \partial\mathcal{S} \tag{8.65}$$

$$\mathbf{S} = [S_{li}] = \left[s_l(\mathbf{x}_{0i})\right], \mathbf{x}_{0i} \in \mathcal{S} \tag{8.66}$$

and

$$\mathbf{A} = \mathbf{S}\mathbf{S}^T = \mathbf{A}^T, \tag{8.67}$$

$$\mathbf{B} = \mathbf{S}\mathbf{G}^T, \tag{8.68}$$

$$\mathbf{R}_b = \mathbf{G}_b\mathbf{G}_b^T = \mathbf{R}_b^T, \tag{8.69}$$

$$\mathbf{b}(k) = \mathbf{B}\mathbf{w}(k-1), \tag{8.70}$$

$$d(k-1) = \frac{1}{2}\mathbf{w}^T(k-1)\mathbf{G}\mathbf{G}^T\mathbf{w}(k-1). \tag{8.71}$$

Hence, $J_s(\mathbf{w}(k), \lambda(k))$ can be written in the following form.

$$\begin{aligned} J_s(\mathbf{w}(k), \lambda(k)) = &\frac{1}{2}\mathbf{w}^T(\mathbf{A} + \lambda(k)\mathbf{R}_b)\mathbf{w}(k) \\ &+ \mathbf{w}^T\mathbf{b}(k) + d(k-1). \end{aligned} \tag{8.72}$$

The necessary conditions for minimizing J_s with respect to the vector $\mathbf{w}(k)$ and $\lambda(k)$ read

$$\frac{\partial J_s}{\partial \mathbf{w}(k)} = (\mathbf{A} + \lambda(k)\mathbf{R}_b)\mathbf{w}(k) + \mathbf{b}(k) = 0, \tag{8.73}$$

$$\frac{\partial J_s}{\partial \lambda(k)} = \frac{1}{2}\mathbf{w}^T\mathbf{R}_b\mathbf{w}(k) = 0. \tag{8.74}$$

8.12.2 Solution in Null Space

Equation (8.74) suggests that the optimal solution $\mathbf{w}(k)$ must reside in the null space of the matrix $\mathbf{R}_b$. Let $\mathbf{Z} \in \mathbb{R}^{N_G \times N_z}$ denote the null space of $\mathbf{R}_b$, where N_z is the number of independent vectors in the null space. $\mathbf{Z}$ therefore consists of N_z column vectors in $\mathbb{R}^{N_G \times 1}$. The solution $\mathbf{w}(k)$ can be expressed as a linear combination of these vectors.

$$\mathbf{w}(k) = \mathbf{Z}\mathbf{v}(k), \tag{8.75}$$

where $\mathbf{v}(k) \in \mathbb{R}^{N_z \times 1}$ is a vector of expansion coefficients. Note that in general, $N_z < N_G$. There is a potential to take less time to compute $\mathbf{w}(k)$ if N_z is significantly less than N_G.

Substituting the solution in Eqs. (8.75)–(8.73) and making use of the property of the null space $\mathbf{R}_b\mathbf{Z} = \mathbf{0}$, we have

$$\mathbf{AZv}(k) + \mathbf{BZv}(k-1) = 0, \tag{8.76}$$

where we have used $\mathbf{b}(k) = \mathbf{BZv}(k-1)$. Pre-multiplying the equation by $\mathbf{Z}^T$, we obtain the coefficient vector as

$$\mathbf{v}(k) = -(\mathbf{Z}^T\mathbf{AZ})^{-1}(\mathbf{Z}^T\mathbf{BZ})\mathbf{v}(k-1) \equiv \mathbf{Cv}(k-1), \tag{8.77}$$

where $\mathbf{C} = -(\mathbf{Z}^T\mathbf{AZ})^{-1}(\mathbf{Z}^T\mathbf{BZ})$.

This solution can be obtained in another way. Since the solution given by Eq. (8.75) satisfies the boundary condition, we don't need to apply the method of Lagrange multiplier. Substituting the solution in Eqs. (8.75)–(8.72), we obtain the error term J_s as,

$$\begin{aligned} J_s(\mathbf{w}(k)) = & \frac{1}{2}\mathbf{v}^T(k)\mathbf{Z}^T\mathbf{AZv}(k) \\ & + \mathbf{v}^T(k)\mathbf{Z}^T\mathbf{BZv}(k-1) + d(k-1). \end{aligned} \tag{8.78}$$

Minimizing J_s with respect to the expansion coefficient vector $\mathbf{v}(k)$ leads to the necessary condition as

$$\frac{\partial J_s}{\partial \mathbf{v}(k)} = \mathbf{Z}^T\mathbf{AZv}(k) + \mathbf{Z}^T\mathbf{BZv}(k-1) = 0. \tag{8.79}$$

This gives the same solution as in Eq. (8.77). Hence, the solution in Eq. (8.77) is in the sense of least square that minimizes $J_s(\mathbf{w}(k))$.

With the help of the concept of null space, we can obtain the RBFNN representation of the initial reliability function in Eq. (8.51) while imposing the boundary condition in Eq. (8.52) at the same time,

$$\begin{aligned} J_0 &= \sum_{i=1}^{N_s} \frac{1}{2}(R(\mathbf{x}_{0i}, \mathbf{w}(0)) - R(\mathbf{x}_{0i}))^2 \quad \mathbf{x}_{0i} \in \mathcal{S} \\ &= \frac{1}{2}\mathbf{v}(0)^T\mathbf{Z}^T\mathbf{GG}^T\mathbf{Zv}(0) - \mathbf{v}(0)^T\mathbf{Z}^T\mathbf{Ge} + \frac{1}{2}\mathbf{e}^T\mathbf{e}, \end{aligned} \tag{8.80}$$

where $R(\mathbf{x}_{0i})$ is the initial distribution of reliability in the safe domain. In the first-passage problems, $R(\mathbf{x}_{0i}) = 1$. $\mathbf{e} \in \mathbb{R}^{N_s\times 1}$ and each element $e_j = 1$ in $\mathbf{e}$. The expansion coefficients $\mathbf{v}(0)$ can be determined by minimizing J_0 in Eq. (8.80).

Remarks

The invertibility of the matrix $\mathbf{Z}^T\mathbf{AZ}$ in Eq. (8.77) is determined by the rank of the matrix $\mathbf{A}$, which heavily depends on the data matrix $\mathbf{S}$. As is well-known in the regression research [12], the data matrix also determines the overall performance of the RBFNN solution in Eq. (8.53). It can be shown through numerical experiments that when a sufficient number of data points $\mathbf{x}_{0i}$ are randomly sampled leading to at least N_G independent algebraic equations to determine the model coefficients $w_j(k)$ at each step k, the inversion of $\mathbf{A}$ will exist and the resulting solution is adequately accurate in the least mean square sense.

Another important remark is that the matrix $\mathbf{C}$ of Eq. (8.77) in the current approach is not a function of time. It can be computed once for all at the beginning. This substantially reduces the computational time for transient solutions.

8.13 First-Passage Examples

8.13.1 Linear Oscillator

Consider a linear oscillator defined by

$$\begin{aligned} dX_1 &= X_2 dt, \\ dX_2 &= (-X_1 - 2\zeta X_2)dt + \sqrt{2\zeta}dB(t), \end{aligned} \tag{8.81}$$

where $\zeta = 0.08$ is used in the numerical results and $B(t)$ is a unit strength Brownian motion such that it has a zero mean value $E[B(t)] = 0$ and a covariance function $E[B(t)B(s)] = \min(t, s)$. Hence, we have the coefficients for the backward Kolmogorov equation as,

$$\begin{aligned} A_1(\mathbf{x}_0, t_0) &= x_{02}, \\ A_2(\mathbf{x}_0, t_0) &= -x_{01} - 2\zeta x_{02}, \\ B_{22}(\mathbf{x}_0, t_0) &= 4\zeta. \end{aligned} \tag{8.82}$$

In this and other examples, we have scaled the linear or linearized system such that the steady state standard deviation of x_1 and x_2 is equal to unity whenever possible. The barrier levels of three types of the safe domain denoted by x_B, x_D and r_E are expressed in terms of the steady state standard deviation σ_x. For example, $x_B = b\sigma_{x_1}$ and b represents the barrier level with respect to σ_{x_1}. In the computation, we have picked $b = 2$.

The domain of computation is selected to be $[-3b, b] \times [-3b, 3b]$ for type B domain, $[-b, b] \times [-3b, 3b]$ for type D domain, and $[-b, b] \times [-b, b]$ for type E domain. The grid size for the mean of Gaussian functions is $[1/16, 1/16]$. The grid size for integration sampling points is $[1/24, 1/24]$.

We use Monte Carlo simulations to check the accuracy of the RBFNN solution. 2×10^5 trajectories of with the length up to 20 periods of the linear oscillator are generated to estimate the first-passage time probability. The period of the oscillator

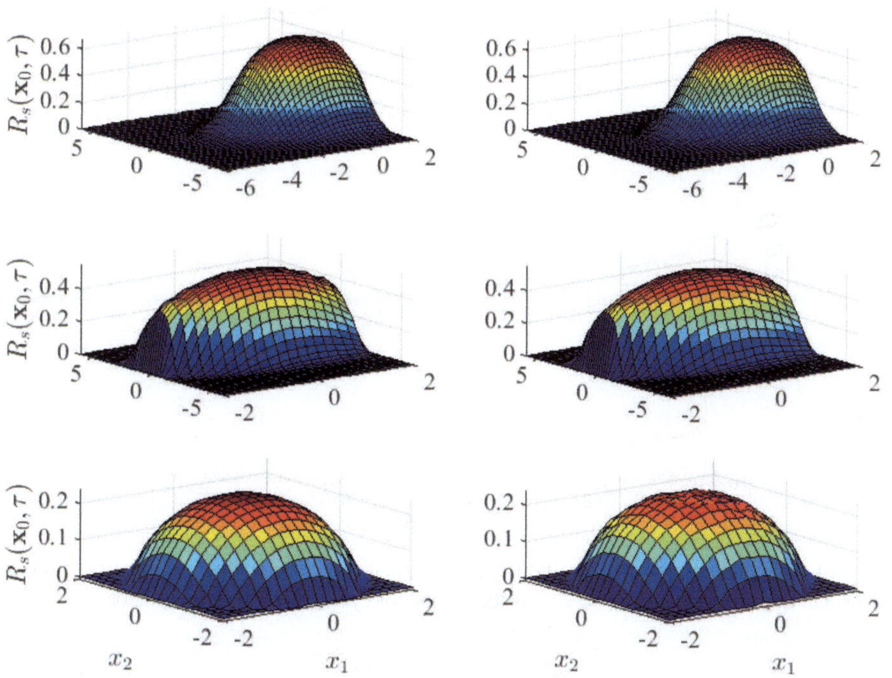

Fig. 8.13 Distribution of the reliability function $R(\mathbf{x}_0, 5T)$ in the safe domain at time $5T$ of system (8.81). Left column: the RBFNN method. Right column: Monte Carlo simulations. First row: Type B. Second row: Type D. Third row: Type E

is $T = 2\pi$. Type E domain can have more crossing events than type B and type D domains, and hence, should have a shorter time step than that for type B and type D domains. For both the RBFNN method and Monte Carlo simulations, we have chosen the same integration time step for comparison purpose. For type B and type D domains, the time step $\Delta\tau = T/200$. For type E domain, $\Delta\tau = T/5000$.

Figure 8.13 shows the distribution of reliability $R(\mathbf{x}_0, 5T)$ at time $5T$ for all the possible initial conditions in the safe domain. The RBFNN solutions for all three domains are in excellent agreement with Monte Carlo simulations.

To quantify the comparison, we define the following average rms error over the safe domain $\mathcal{S}$.

$$\Delta R(\tau) = \sqrt{\frac{1}{m(\mathcal{S})}\int_{\mathcal{S}}(R(\mathbf{x}_0, \mathbf{w}(k)) - R^{MC}(\mathbf{x}_0, \tau))^2 d\mathbf{x}_0} \tag{8.83}$$

where $\tau = k\Delta\tau$ and $m(\mathcal{S})$ denotes a measure of the safe domain $\mathcal{S}$, which can be the length of an interval, the area, or the volume, *etc.* $R^{MC}(\mathbf{x}_0, \tau)$ is the reliability function computed by Monte Carlo simulations at the kth time step starting from the

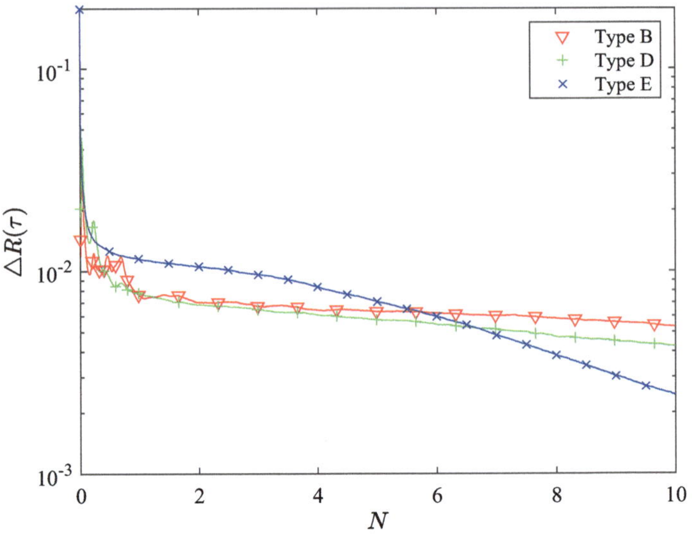

Fig. 8.14 The rms error of the reliability function obtained by the RBFNN method as compared to Monte Carlo simulations. $N = \tau/T$ is the number of cycles of the linear oscillator

initial condition $\mathbf{x}_0$. It should be noted that because the reliability function does not satisfy the normalization condition except at $\tau = 0$, the rms error is not normalized.

Figure 8.14 shows the rms errors for three types of the safe domain. The integration for the rms error is carried over the domain $[-b, b] \times [-b, b]$ because the reliability function value is nearly zero outside this domain for $k > 0$. It should be pointed out that the rms error for all three types of domain decreases from its initial relatively large value to a value below 0.02 in less than 1/5 period of the linear oscillator.

To further illustrate the utility of the RBFNN method, we present several examples of the first-passage time probability and its density function for different initial conditions. The RBFNN results are compared with the same Monte Carlo simulations discussed earlier. Figure 8.15 shows the first-passage time probability and its density function for three types of the safe domain when the system starts from the initial condition $\mathbf{x}_0 = (0, 0)$. Figure 8.16 shows the same results for the initial condition $\mathbf{x}_0 = (1, 1)$. Because the point $\mathbf{x}_0 = (1, 1)$ is closer to the boundary of the safe domain. The results in Fig. 8.16 indicate much higher probability and its density at earlier time than those for the initial condition $\mathbf{x}_0 = (0, 0)$ in Fig. 8.15.

The RBFNN method is quite efficient and can be used to study the effect of various parameters on the first-passage time probability and its density function.

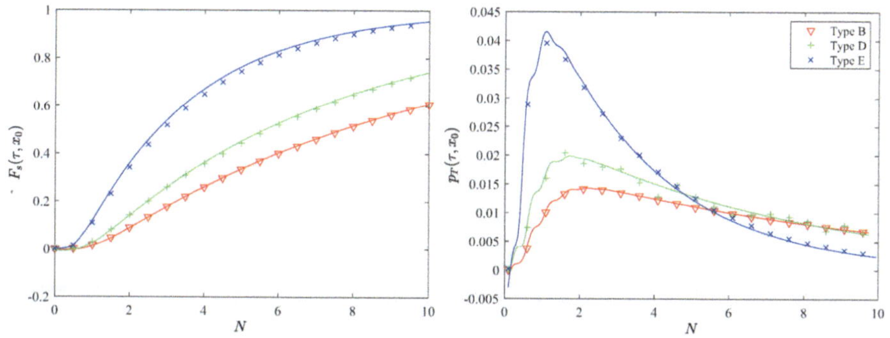

Fig. 8.15 The first-passage time probability and its density function of the linear system starting from the initial condition $\mathbf{x}_0 = (0, 0)$. RBFNN solution: solid lines. Monte Carlo simulations: $\triangledown$, Type B; $+$, Type D; $\times$, Type E

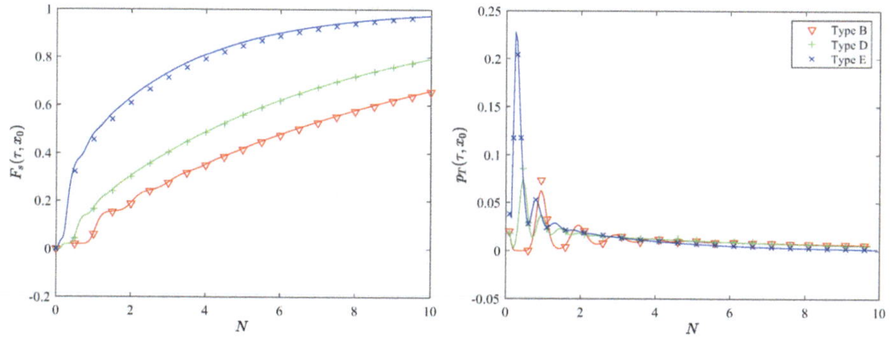

Fig. 8.16 The first-passage time probability and its density function of the linear system starting from the initial condition $\mathbf{x}_0 = (1, 1)$. RBFNN solution: solid lines. Monte Carlo simulations: $\triangledown$, Type B; $+$, Type D; $\times$, Type E

8.13.2 Van der Pol Oscillator

We consider a much more complex nonlinear stochastic system, i.e. the Van der Pol oscillator under the Gaussian white noise excitation.

$$\ddot{x} + 2\zeta(\varepsilon x^2 - 1)\dot{x} + x = \sqrt{2\zeta}w(t), \tag{8.84}$$

where $w(t)$ is a unit Gaussian white noise such that $E[w(t)] = 0$ and $E[w(t)w(t + \tau)] = 2\delta(\tau)$. We first consider the parameters $b = 2.0$, $\zeta = 0.08$ and $\varepsilon = 0.5$. The parameter $\varepsilon = 0.5$ indicates that the system is strongly nonlinear. Figures 8.17 and 8.18 show the first-passage time probability and its density function of three types of the safe domain for the initial conditions $\mathbf{x}_0 = (0, 0)$ and $\mathbf{x}_0 = (1, 1)$. As compared to the linear oscillator, the strongly nonlinear Van der Pol oscillator tends to have larger excursions in the state space and has more and earlier crossings of the boundary of the

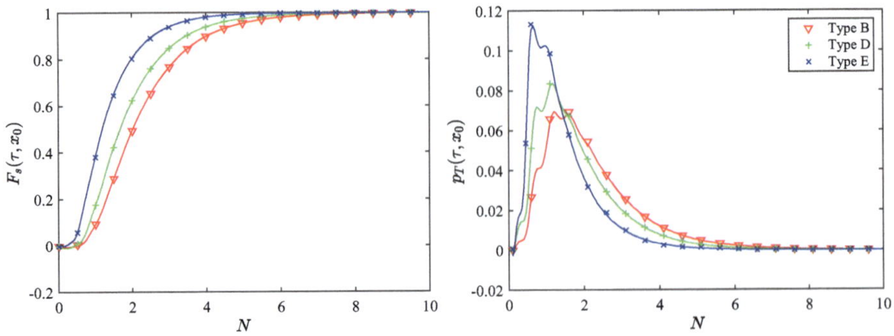

Fig. 8.17 The first-passage time probability and its density function of the Van der Pol system starting from the initial condition $\mathbf{x}_0 = (0, 0)$. RBFNN solution: solid lines. Monte Carlo simulations: $\triangledown$, Type B; $+$, Type D; $\times$, Type E

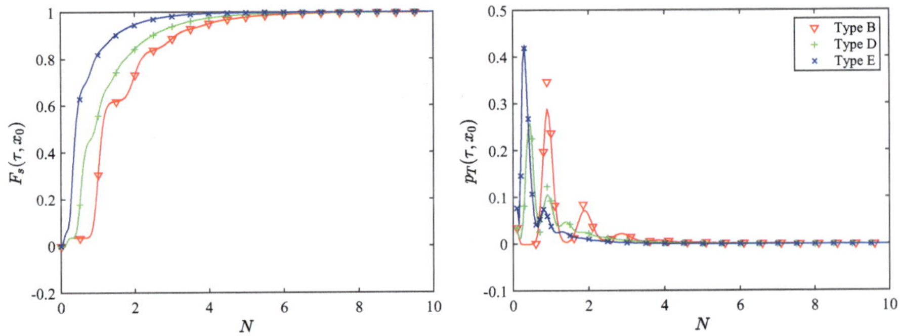

Fig. 8.18 The first-passage time probability and its density function of the Van der Pol system starting from the initial condition $\mathbf{x}_0 = (1, 1)$. RBFNN solution: solid lines. Monte Carlo simulations: $\triangledown$, Type B; $+$, Type D; $\times$, Type E

safe domain. As is the case of the linear oscillator, the initial condition $\mathbf{x}_0 = (1, 1)$ is closer to the boundary and leads to higher first-passage time probability than $\mathbf{x}_0 = (0, 0)$.

To illustrate the ability of the RBFNN method to deal with strongly nonlinear systems, we consider the parameters $b = 2.0$, $\zeta = 0.5$ and $\varepsilon = 2.0$, which lead to a Van der Pol oscillator with strong nonlinear damping. The response of the oscillator is highly distorted in the state space [10]. Figures 8.19 and 8.20 show the first-passage time probability and its density function of three types of the safe domain starting from the initial conditions $\mathbf{x}_0 = (0, 0)$ and $\mathbf{x}_0 = (1, 1)$. The RBFNN solutions are again found to be in good agreement with Monte Carlo simulations.

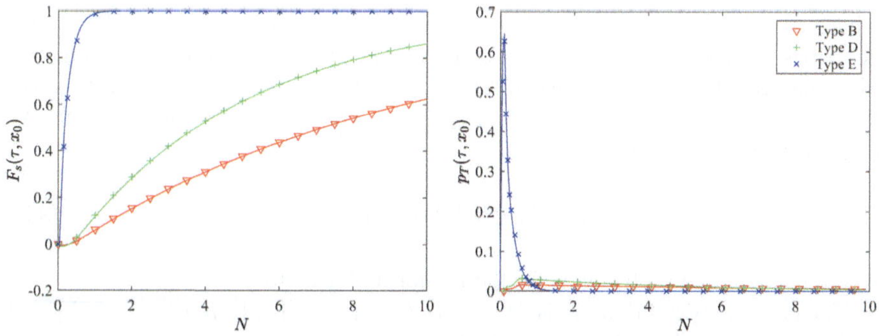

Fig. 8.19 The first-passage time probability and its density function of the Van der Pol system with strong nonlinear damping starting from the initial condition $\mathbf{x}_0 = (0, 0)$. RBFNN solution: solid lines. Monte Carlo simulations: ∇, Type B; $+$, Type D; $\times$, Type E

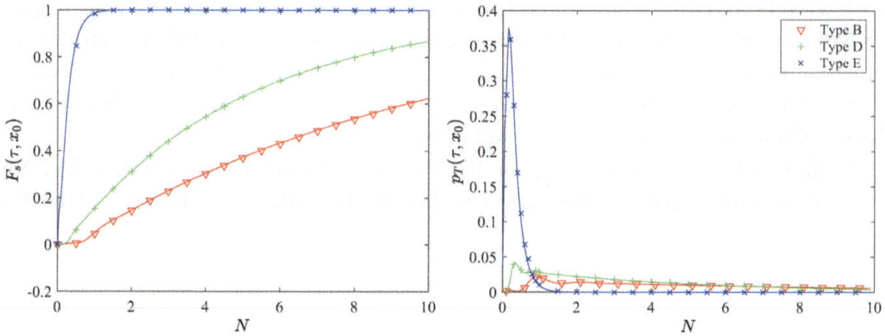

Fig. 8.20 The first-passage time probability and its density function of the Van der Pol system with strong nonlinear damping starting from the initial condition $\mathbf{x}_0 = (1, 1)$. RBFNN solution: solid lines. Monte Carlo simulations: ∇, Type B; $+$, Type D; $\times$, Type E

8.13.3 Duffing System

The advantage of the RBFNN method is its ability to handle multi-degree of freedom dynamic systems. We use a two-degree-of-freedom Duffing system to demonstrate this point.

$$\begin{aligned}
\dot{x}_1 &= x_2, \\
\dot{x}_2 &= -\mu_1 x_2 - \omega_1^2 x_1 - k_1 x_1^3 + \alpha_1 x_3 \\
&\quad + \beta_1 x_1^2 x_3 + \sqrt{2\zeta_1} w_1(t), \\
\dot{x}_3 &= x_4, \\
\dot{x}_4 &= -\mu_2 x_4 - \omega_2^2 x_3 - k_2 x_3^3 + \alpha_2 x_1 \\
&\quad + \beta_2 x_1^3 + \sqrt{2\zeta_2} w_2(t),
\end{aligned} \tag{8.85}$$

where $w_1(t)$ and $w_2(t)$ are independent unit Gaussian white noises. This system describes nonlinear ship motion in random seas [13]. In the following numerical example, we take the parameters $\mu_1 = 0.2$, $\mu_2 = 0.2$, $\omega_1 = 0.2$, $\omega_2 = 0.4$, $k_1 = 0.3$, $k_2 = 0.5$, $\alpha_1 = 0.4$, $\alpha_2 = 0.2$, $\beta_1 = 0.12$, $\beta_2 = 0.08$, $\zeta_1 = 0.04$ and $\zeta_2 = 0.04$. We consider type E domain only defined as

$$\mathcal{S} = \left[\mathbf{x} \left| \mathbf{x} \in \mathbb{R}^{4\times 1} \text{ and } ||\mathbf{x}|| \leq b \right.\right], \tag{8.86}$$

where $b = 1.5$ and $|| \cdot ||$ stands for the L_2 norm of a vector.

The computational domain is $D = [-b, b] \times [-b, b] \times [-b, b] \times [-b, b]$. D is divided into $21 \times 21 \times 21 \times 21$ grids. Each grid point $\mathbf{x}_j$ defines the mean $\boldsymbol{\mu}_j$ of a Gaussian function with standard deviation $\boldsymbol{\sigma}_j = 0.8\mathbf{h}_j$, where $\mathbf{h}_j$ is a vector of grid sizes. We uniformly sample $23 \times 23 \times 23 \times 23$ points to compute the data matrices. We select $11 \times 11 \times 11 \times 11$ initial conditions in the region $[-1, 1] \times [-1, 1] \times [-1, 1] \times [-1, 1]$ to compute the rms error of the RBFNN solutions with respect to Monte Carlo simulations. In Monte Carlo simulations, 2×10^5 trajectories are generated starting from each initial condition. The total integration time is $5T$ where $T = 2\pi$. For the RBFNN method, the time interval T is divided into 5000 steps. For Monte Carlo simulations, T is divided into 5000 steps.

Figure 8.21 shows the rms error of the reliability function obtained by the RBFNN method as compared to Monte Carlo simulations. It is interesting to observe that the

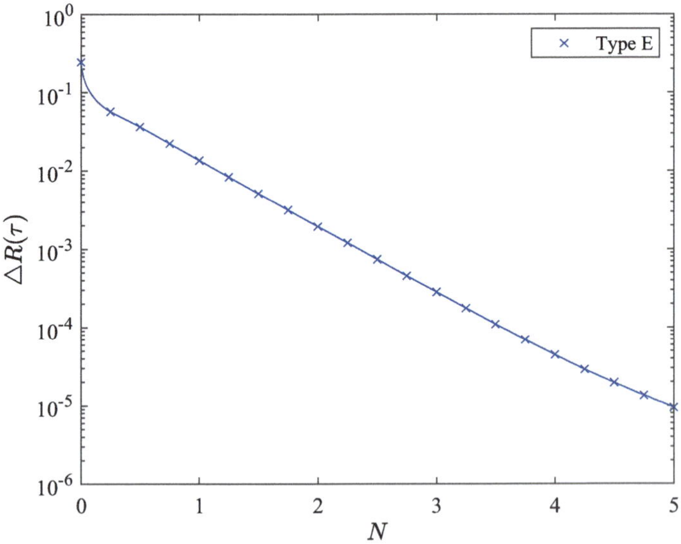

Fig. 8.21 The rms error of the reliability function obtained by the RBFNN method as compared to Monte Carlo simulations for the two-degree-of-freedom example in Eq. (8.85). $N = \tau/T$ and $T = 2\pi$

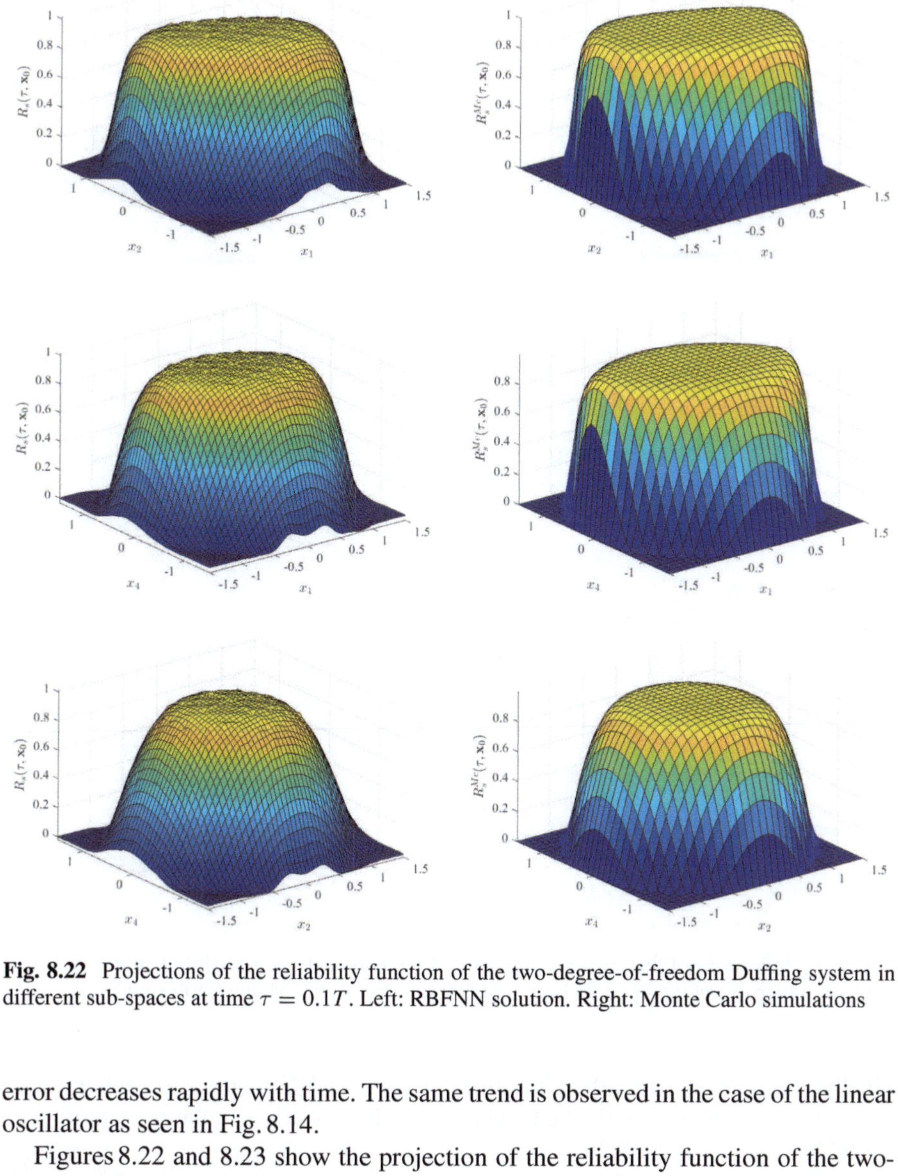

Fig. 8.22 Projections of the reliability function of the two-degree-of-freedom Duffing system in different sub-spaces at time $\tau = 0.1T$. Left: RBFNN solution. Right: Monte Carlo simulations

error decreases rapidly with time. The same trend is observed in the case of the linear oscillator as seen in Fig. 8.14.

Figures 8.22 and 8.23 show the projection of the reliability function of the two-degree-of-freedom Duffing system in different sub-spaces at time $\tau = 0.1T$ and $\tau = 1T$ where $T = 2\pi$. Such accurate results for two-degree-of-freedom dynamic systems are rare in the literature.

Next, we present the results of the first-passage time probability and its density function of the two-degree-of-freedom Duffing system starting from the initial conditions $\mathbf{x}_0 = (0, 0, 0, 0)$ and $\mathbf{x}_0 = (0.4, 0.4, 0.4, 0.4)$. The results obtained by the RBFNN method are in good agreement with those of Monte Carlo simulations.

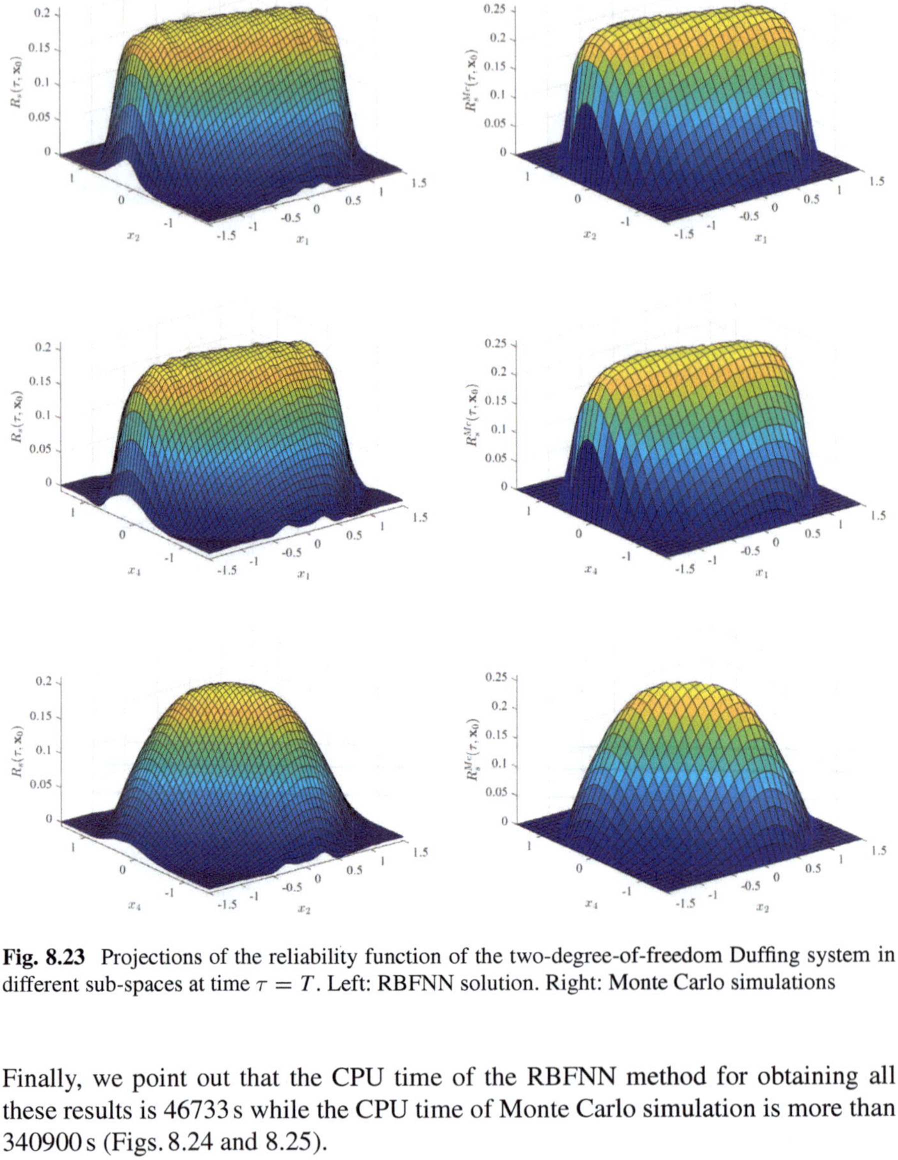

Fig. 8.23 Projections of the reliability function of the two-degree-of-freedom Duffing system in different sub-spaces at time $\tau = T$. Left: RBFNN solution. Right: Monte Carlo simulations

Finally, we point out that the CPU time of the RBFNN method for obtaining all these results is 46733 s while the CPU time of Monte Carlo simulation is more than 340900 s (Figs. 8.24 and 8.25).

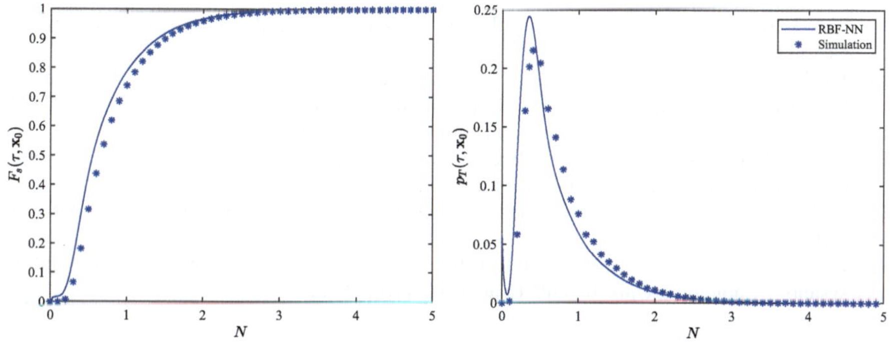

Fig. 8.24 The first-passage time probability and its density function of the two-degree-of-freedom Duffing system starting from the initial condition $\mathbf{x}_0 = (0, 0, 0, 0)$. RBFNN solution: solid line. Monte Carlo simulations: *

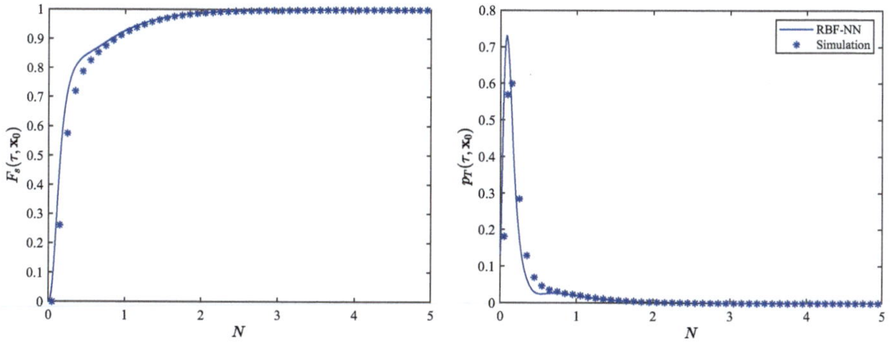

Fig. 8.25 The first-passage time probability and its density function of the two-degree-of-freedom Duffing system starting from the initial condition $\mathbf{x}_0 = (0.4, 0.4, 0.4, 0.4)$. RBFNN solution: solid line. Monte Carlo simulations: *

References

1. Wang X, Jiang J, Hong L, Sun JQ (2022) Random vibration analysis with radial basis function neural networks. Int J Dyn Control 10(5):1385–1394. https://doi.org/10.1007/s40435-021-00893-2
2. Wang X, Jiang J, Hong L, Sun JQ (2022) First-passage problem in random vibrations with radial basis function neural networks. J Vib Acoust 144(5):051,014. https://doi.org/10.1115/1.4054437
3. Han J, Jentzen A, E W (2018) Solving high-dimensional partial differential equations using deep learning. PNAS 115(34):8505–8510
4. Sirignano J, Spiliopoulos K (2018) DGM: a deep learning algorithm for solving partial differential equations. arXiv:1708.07469v5 [q-fin.MF]
5. Muller J, Zeinhofer M (2020) Deep Ritz revisited. arXiv:1912.03937v2 [math.NA]
6. He J, Li L, Xu J, Zheng C (2018) ReLU deep neural networks and linear finite elements. arXiv:1807.03973v2 [math.NA]
7. Lagaris IE, Likas AC, Papageorgiou DG (2000) Neural-network methods for boundary value problems with irregular boundaries. IEEE Trans Neural Netw 11(5):1041–1049

8. Park J, Sandberg IW (1991) Universal approximation using radial-basis-function networks. Neural Comput 3:246–257
9. Wu Y, Wang H, Zhang B, Du KL (2012) Using radial basis function networks for function approximation and classification. ISRN Appl Math 1–34. https://doi.org/10.5402/2012/324194
10. Sun JQ (2006) Stochastic dynamics and control. Elsevier Science Ltd., Oxford, U.K.
11. Sun JQ, Hsu CS (1990) The generalized cell mapping method in nonlinear random vibration based upon short-time Gaussian approximation. J Appl Mech 57(4):1018–1025
12. Nelles O (2001) Nonlinear system identification - from classical approaches to neural networks and fuzzy models. Springer, Berlin
13. Cao ZB, Le Y (2015) Analysis of bifurcation and chaos in a two-degree-of-freedom duffing system. J Chongqing Univ Technol (Nat Sci) 29(10):79–82

Chapter 9
Method of Feynman-Kac

9.1 Background

In Chap. 6, we have discussed two approaches to derive the FPK equation governing the probability density function of the response of stochastic differential equation. The FPK equation is a parabolic partial differential equation (PDF), which is widely employed to describe the evolution of complex systems in various fields, including physics, engineering and finance. The FPK equation is just one of many examples in Physics and Mathematics that suggest that there is a relationship between certain PDEs and stochastic differential equations. The Feynman-Kac formula provides a connection between a semilinear parabolic PDE and a stochastic process with the martingale property [1, 2].

The solution to the parabolic PDE is proven to be the expectation of the martingale process driven by a Brownian motion. The theory of backward stochastic differential equations (BSDEs) extends the application of the Feynman-Kac method from semilinear parabolic PDEs to fully nonlinear PDEs [3, 4]. Xu further applied the concepts of the Feynman-Kac formula and BSDE to the complex PDEs [5]. The Feynman-Kac formula provides a route to solve PDEs with the help of simulated random paths without the need for derivative calculations of the differential operator. An algorithm of variational quantum imaginary time evolution has been developed to solve the PDE in order to obtain the expectations of the SDEs [6].

The primary application of the Feynman-Kac formula remains providing a method for solving PDEs by simulating stochastic processes, thereby circumventing the need for derivative calculations. Monte Carlo simulations (MCS) can be employed to directly generate random trajectories in order to obtain the expectation of the stochastic process, which represents the solution to the PDE. However, direct simulations would require significant computational efforts [7–9].

To enhance the computational efficiency, the h-conditioned Green's function, which enables the representation of the integral along random paths as a volume integral, is employed instead of directly simulating random trajectories to obtain the

J.-Q. Sun, *Random Vibration with Machine Learning Method*,
Nonlinear Physical Science, https://doi.org/10.1007/978-981-95-0812-9_9

expectation of the stochastic process [10]. Dalang et al. established a probabilistic representation for PDEs that is different from the Feynman-Kac formula. The advantage of their method lies in deriving an analytical expression for the moments of the response process without the need for numerical simulations [11]. Computation of the expectation of the martingale process with direct simulations can be even more intensive for high dimensional systems [7, 8].

Deep learning methods based on variational formulations for the associated stochastic differential equations are investigated extensively in [12]. The Feynman-Kac formula is also used in the study, leading to a very efficient computational algorithm.

Furthermore, the derivative-free deep neural network method has been studied to solve viscous incompressible fluid flow problems [13] and elliptical PDEs [14]. Weinan et al. solve a class of parabolic PDEs by reformulating them as BSDEs using deep neural networks [15].

Due to their robust fitting capability and adaptability to complex data relationships, artificial neural networks (ANNs) have been employed as a universal method to solve PDEs [15–17]. This leads to physics informed neural networks commonly known as PINN. ANNs transform the PDE solution process into optimization problems, whose objective is to minimize a suitably defined loss function, such that the network output approximates the solution to a high level of accuracy. The simplicity and flexibility of the PINN approach render it suitable for a wide range of PDEs. Some literature on solving the FPK equation using the PINN approach includes [18–21].

The loss function of PINN usually consists of residuals of the PDE, boundary conditions and possibly other constraints and therefore requires the computation of high order derivatives of the equation. The computation of derivatives in the loss function in the repeated trainings of the neural networks can substantially increase the CPU time and sacrifice the accuracy of the solution. A Monte Carlo method was proposed to compute second derivatives, which proves to be time-consuming [22]. Furthermore, this approach particularly with deep neural networks may also face challenges such as vanishing gradient or exploding gradient problems, which can impede the convergence and overall effectiveness [23].

Although the calculation of derivatives can be circumvented by the Feynman-Kac formula, the traditional neural network methods still require numerical simulations to compute the expectation of random paths of the martingale stochastic process, which remains time-consuming and inconvenient.

In this chapter, we present an effective way to compute the expectation of the martingale process by making use of the short-time Gaussian approximation of the response process. We also utilize the RBFNN to accurately approximate the PDF of the response process. RBFNN is an example of PINN and has been successfully applied to obtain transient and stationary PDFs of the FPK equation of nonlinear systems [24–26] and the reliability function of the first-passage problem [27]. Since the output of the RBFNN is a weighted sum of Gaussian activation functions, by applying the short-time Gaussian approximation (STGA) technique, we can derive the analytical expression for the expectation of the Feynman-Kac formula. This approach obviates the need for numerical simulations.

Compared to directly solving PDEs using RBFNN, the Feynman-Kac method additionally eliminates the need for derivative calculations, thereby reducing computational time. Numerical examples, including a strongly nonlinear system, a non-smooth system, and a high-dimensional system, are employed to demonstrate the feasibility and effectiveness of the Feynman-Kac method.

Section 9.2 provides a brief description of the Feynman-Kac formula and the FPK equation. In Sect. 9.3, we elaborate on the implementation of the Feynman-Kac formula utilizing the RBFNN approach. We apply the STGA technique to derive the analytical expression for the expectation of the stochastic process, thereby circumventing both derivative calculations and numerical simulations. In Sect. 9.4, we briefly review a RBFNN approach of directly solving the FPK equation. In Sect. 9.5, three examples are presented, including a strongly nonlinear system, a non-smooth tri-stable system, and a high-dimensional system, to validate the feasibility and effectiveness of the Feynman-Kac method. The computational time of the Feynman-Kac method is compared with that of the equation-based RBFNN method in Sect. 9.7.

9.2 Feynman-Kac Formula

We first present an overview of the Feynman-Kac formula, which provides a connection between a second-order parabolic PDE and a stochastic process with the martingale property. However, we shall not show how to derive the FPK equation from the Feynman-Kac formula. The interested reader should refer to the reference [6].

Consider the following n-dimensional parabolic PDE,

$$\frac{\partial p(\mathbf{x}, t)}{\partial t} + \mathcal{L}_t[p(\mathbf{x}, t)] - c(\mathbf{x}, t)p(\mathbf{x}, t) = 0, \tag{9.1}$$

where the differential operator $\mathcal{L}_t[p]$ is defined as,

$$\mathcal{L}_t[p] \equiv \sum_{i=1}^{n} a_i(\mathbf{x}, t)\frac{\partial p}{\partial x_i} + \frac{1}{2}\sum_{i=1}^{n}\sum_{j=1}^{n} b_{ij}(\mathbf{x}, t)\frac{\partial^2 p}{\partial x_i \partial x_j}. \tag{9.2}$$

The coefficients $a_i(\mathbf{x}, t), b_{ik}(\mathbf{x}, t)$ and $c(\mathbf{x}, t)$ are known functions $\mathbb{R}^n \times [0, T] \to \mathbb{R}$, $\mathbf{a} = [a_i] \in \mathbb{R}^{n\times 1}$ and $\mathbf{b} = [b_{ik}] \in \mathbb{R}^{n\times m}$. As discussed in the introduction [6], the Feynman-Kac formula can be used to derive PDE (9.1) governing the PDF of the n dimensional stochastic process $\mathbf{X}(t) = [X_j(t)] \in \mathbb{R}^n$ satisfying the following Itô differential equation,

$$dX_j = a_j(\mathbf{X}, t)dt + \sum_{k=1}^{m} \sigma_{jk}(\mathbf{X}, t)dB_k(t), \tag{9.3}$$

where $b_{ij}(\mathbf{x}, t) = \sum_{k=1}^{m} \sigma_{ik}\sigma_{jk}(\mathbf{x}, t)$ and $B_j(t)$ $(j = 1, 2, \ldots, m)$ are independent unit Brownian motions. Hence, the FPK equation for the stochastic process $\mathbf{X}(t)$ satisfying Eq. (9.3) must be a special case of Eq. (9.1).

Consider the n-dimensional FPK equation with the initial condition.

$$\frac{\partial p(\mathbf{x}, t)}{\partial t} = \mathcal{L}_{FPK}[p(\mathbf{x}, t)] \equiv -\sum_{i=1}^{n} \frac{\partial[m_i(\mathbf{x})p(\mathbf{x}, t)]}{\partial x_i} + \frac{1}{2}\sum_{i=1}^{n}\sum_{j=1}^{n} \frac{\partial^2[b_{ij}(\mathbf{x})p(\mathbf{x}, t)]}{\partial x_i \partial x_j}, \tag{9.4}$$

$$p(\mathbf{x}, 0) = p_0(\mathbf{x}), \tag{9.5}$$

where the deterministic functions $m_i(\mathbf{x})$ and $b_{ij}(\mathbf{x})$ are the drift and diffusion terms. The FPK operator can be expressed in the form of a linear parabolic equation as,

$$\mathcal{L}_{FPK}[p(\mathbf{x}, t)] = \mathcal{L}_t[p(\mathbf{x}, t)] - c(\mathbf{x})p(\mathbf{x}, t), \tag{9.6}$$

where $\mathcal{L}_t[p]$ is defined in Eq. (9.2), and

$$c(\mathbf{x}) = \sum_{i=1}^{n} \frac{\partial m_i(\mathbf{x})}{\partial x_i} - \frac{1}{2}\sum_{i=1}^{n}\sum_{j=1}^{n} \frac{\partial^2 b_{ij}(\mathbf{x})}{\partial x_i \partial x_j},$$
$$a_i(\mathbf{x}) = -m_i(\mathbf{x}) + \sum_{j=1}^{n} \frac{\partial b_{ij}(\mathbf{x})}{\partial x_j}. \tag{9.7}$$

Note that this is a special case when $a_i(\mathbf{x})$, $b_{ij}(\mathbf{x})$ and $c(\mathbf{x})$ are not explicit functions of time.

Consider again the parabolic PDE (9.1) and the stochastic process $\mathbf{X}(t)$, and define another stochastic process $\mathcal{M}(t)$ as,

$$\mathcal{M}(t) = e^{-\int_0^t c(\mathbf{X}(s), s)ds} p(\mathbf{X}, t), \quad 0 < t \leq T, \tag{9.8}$$

where $p : \mathbb{R}^n \times [0, T] \to \mathbb{R}$ is an arbitrary function at this point. The differentiation of $\mathcal{M}(t)$ can be obtained by Itô's lemma as,

$$
\begin{aligned}
d\mathcal{M}(t) &= e^{-\int_0^t c(\mathbf{X}(s),s)ds} dp(\mathbf{X}, t) \\
&\quad - e^{-\int_0^t c(\mathbf{X}(s),s)ds} p(\mathbf{X}, t)c(\mathbf{X}, t)dt \\
&= e^{-\int_0^t c(\mathbf{X}(s),s)ds} \left(\frac{\partial p}{\partial t} + \mathcal{L}_t[p] - cp \right) dt \\
&\quad + \sum_{j=1}^{n} \sum_{k=1}^{m} \sigma_{jk} \frac{\partial p}{\partial x_j} dB_k(t).
\end{aligned} \tag{9.9}
$$

When $p(\mathbf{x}, t)$ is the solution of PDE (9.1), we take the expectation of Eq. (9.9) and obtain

$$
d\mathrm{E}[\mathcal{M}(t)] = 0. \tag{9.10}
$$

Hence, $\mathcal{M}(t)$ is a martingale process. Integrating this equation over $[0, T]$ gives $\mathrm{E}[\mathcal{M}(0)] = \mathrm{E}[\mathcal{M}(T)]$ where $T > 0$ is arbitrary. We have

$$
p(\mathbf{x}, 0) = \mathrm{E}\left[e^{-\int_0^T c(\mathbf{X}(s),s)ds} p(\mathbf{X}(T), T) | \mathbf{X}(0) = \mathbf{x} \right]. \tag{9.11}
$$

The equation relates the initial distribution $p(\mathbf{x}, 0)$ to the PDF $p(\mathbf{x}, T)$ at an arbitrary time $T > 0$. This is an implicit integral equation for determining the PDF of the stochastic process $\mathbf{X}(t)$ without the differential operations of PDE (9.1). Therefore, the derivative calculations, which are typically time-consuming in computations for solving PDEs, are not required when $p(\mathbf{x}, T)$ is obtained from the integral equation.

Equation (9.11) needs samples of the stochastic process $\mathbf{X}(t)$ over the time interval $[0, T]$ to evaluate the expectation. For nonlinear systems, Monte Carlo simulations have been the only effective method to compute the expectation. In the following, we present a neural networks method to obtain the PDF solution from Eq. (9.11) subject to the constraint of normalization.

$$
\int_{\mathbb{R}^n} p(\mathbf{x}) d\mathbf{x} = 1. \tag{9.12}
$$

9.3 Solution via Feynman-Kac Formula

The solution method making use of the Feynman-Kac formula consists of a few steps. In the first step, we divide the time interval $[0, T]$ into a collection of small time intervals $[k\Delta t, (k+1)\Delta t]$ $(k = 0, 1, \ldots)$. In each sub-interval, the martingale property of the stochastic process $\mathcal{M}(t)$ holds

$$
\begin{aligned}
p(\mathbf{x}, k\Delta t) = \mathrm{E}\big[& e^{-\int_{k\Delta t}^{(k+1)\Delta t} c(\mathbf{X}(s),s)ds} \times \\
& p(\mathbf{X}((k+1)\Delta t), (k+1)\Delta t) | \mathbf{X}(k\Delta t) = \mathbf{x} \big].
\end{aligned} \tag{9.13}
$$

In the second step, we develop a way to compute the conditional expectation on the right hand side of Eq. (9.13). $p(\mathbf{x}, (k+1)\Delta t)$ is a short-time response starting from the deterministic initial condition $p(\mathbf{x}, k\Delta t)$. Because the stochastic process $\mathbf{X}(t)$ satisfies Eq. (9.3), the STGA of the PDF of the response $\mathbf{X}((k+1)\Delta t)$ is valid. The details of the STGA strategy can be found in [28]. The mathematical expression of the STGA approximation of the PDF will be presented later.

With the PDF of the response $\mathbf{X}((k+1)\Delta t)$, we can analytically evaluate the conditional expectation of the nonlinear and unknown function of $\mathbf{X}(t)$ in Eq. (9.13).

In the third step, we propose a neural networks representation of the unknown function $p(\mathbf{x}, (k+1)\Delta t)$, which is a probability density function. From our previous experience [24–27], we find that the radial basis function neural networks (RBFNN) with Gaussian activation functions can be an excellent choice for modeling $p(\mathbf{x}, (k+1)\Delta t)$.

9.3.1 RBFNN Solution

Let $\bar{p}(\mathbf{x}, \mathbf{w}(k))$ be the function approximating $p(\mathbf{x}, k\Delta t)$ at the kth time step where $\mathbf{w}(k)$ denotes all the trainable coefficients of the neural networks. The trial solution $\bar{p}(\mathbf{x}, \mathbf{w}(k))$ is written as the RBFNN with Gaussian activation functions

$$\bar{p}(\mathbf{x}, \mathbf{w}(k)) = \sum_{j=1}^{N_G} w_j(k) G_j(\mathbf{x}), \tag{9.14}$$

where N_G represents the total number of neurons. $G_j(\mathbf{x}) \equiv G(\mathbf{x}, \boldsymbol{\mu}_j, \boldsymbol{\Sigma}_j)$ is a n-variate Gaussian radial basis function with the mean $\boldsymbol{\mu}_j$ and covariance matrix $\boldsymbol{\Sigma}_j = \mathrm{diag}[\boldsymbol{\sigma}_j^2]$. The n-variate Gaussian radial basis function is separable and can be expressed as a product of multiple uni-variate Gaussian functions as,

$$G(\mathbf{x}, \boldsymbol{\mu}_j, \boldsymbol{\Sigma}_j) = \prod_{k=1}^{n} g(x_k, \mu_{jk}, \sigma_{jk}), \tag{9.15}$$

$$g(x_k, \mu_{jk}, \sigma_{jk}) = \frac{1}{\sqrt{2\pi\sigma_{jk}^2}} \exp\left[-\frac{1}{2\sigma_{jk}^2}(x_k - \mu_{jk})^2\right],$$

where μ_{jk} is the kth component of the mean $\boldsymbol{\mu}_j$ and $\sigma_{j,k}$ is the kth component of the standard deviation $\boldsymbol{\Sigma}_j$. In this chapter, the means $\boldsymbol{\mu}_j$ and standard deviations $\boldsymbol{\sigma}_j$ are taken as constants.

The RBFNN solution (9.14) is a neural network with a single hidden layer using the Gaussian activation function.

This shallow neural network has been proven to possess universal approximation capabilities [29]. Furthermore, the normalization condition (9.12) now reads,

$$\sum_{j=1}^{N_G} w_j(k) = 1, \quad k \geq 0. \tag{9.16}$$

Let D_G denote the domain where the system stays with probability not less than 99%. We discretize the domain into uniform grids, which represent the means of Gaussian neurons. The standard deviations for all the Gaussian function are set equal to the grid size.

9.3.2 Loss Function

Let us consider the time-invariant case. Assume that the duration of each sub-interval Δt is sufficiently small. Let

$$\int_{(k-1)\Delta t}^{k\Delta t} c(\mathbf{X}(s))ds = \bar{c}\Delta t, \tag{9.17}$$

where $\bar{c}$ is determined from the trapezoidal rule. Before presenting the loss function, we introduce a shorthand for the expectation with the help of Eq. (9.14)

$$\begin{aligned} &\mathrm{E}\left[e^{-\bar{c}\Delta t}\bar{p}(\mathbf{X}(k\Delta t), \mathbf{w}(k))|\mathbf{X}((k-1)\Delta t) = \mathbf{x}_i\right] \\ &= \mathrm{E}\left[e^{-\bar{c}\Delta t}\sum_{j=1}^{N_G} w_j(k)G_j(\mathbf{X}(k\Delta t))|\mathbf{X}((k-1)\Delta t) = \mathbf{x}_i\right], \end{aligned} \tag{9.18}$$

$$e_j(\mathbf{x}_i) = \mathrm{E}\left[e^{-\bar{c}\Delta t}G_j(\mathbf{X}(k\Delta t))|\mathbf{X}((k-1)\Delta t) = \mathbf{x}_i\right]. \tag{9.19}$$

where $e_j(\mathbf{x}_i)$ is the expectation of the Gaussian activation function $G_j(\mathbf{x})$.

Define a loss function and make use of Eq. (9.19). We have

$$\begin{aligned} J(\mathbf{w}(k), k) &= \sum_{i=1}^{N_s} \left[p(\mathbf{x}_i, (k-1)\Delta t) - w_j(k)e_j(\mathbf{x}_i)\right]^2 \\ &= \sum_{i=1}^{N_s} r^2(\mathbf{x}_i), \end{aligned} \tag{9.20}$$

where $r(\mathbf{x}_i)$ measures the degree to which the trial solution satisfies the martingale property at the sampling point x_i,

$$r(\mathbf{x}_i) = p(\mathbf{x}_i, (k-1)\Delta t) - \sum_{j=1}^{N_G} w_j(k)e_j(\mathbf{x}_i). \tag{9.21}$$

9.3.3 STGA Technique

The short-time probability density solution of the stochastic process, as defined in Eq. (9.3), starting from the initial condition $\mathbf{X}(k\Delta t) = \mathbf{x}_i$, is approximately a Gaussian PDF with the mean $\mathbf{A}_i = \mathbf{x}_i + \mathbf{a}(\mathbf{x}_i)\Delta t$ and covariance $\mathbf{B}_i = \boldsymbol{\sigma}\boldsymbol{\sigma}^T(\mathbf{x}_i)\Delta t$, as follows [30],

$$
\begin{aligned}
p(\mathbf{x}, (k+1)\Delta t|\mathbf{x}_i, k\Delta t) &= \frac{1}{\sqrt{(2\pi)^{2n}\det(\mathbf{B}_i)}} \\
&\times \exp\left[-\frac{1}{2}(\mathbf{x}-\mathbf{A}_i)^T\mathbf{B}_i^{-1}(\mathbf{x}-\mathbf{A}_i)\right] \\
&\equiv G(\mathbf{x}, \mathbf{A}_i, \mathbf{B}_i).
\end{aligned}
\tag{9.22}
$$

With this Gaussian PDF, the expectation $\mathrm{E}[G_j(\mathbf{X}(k\Delta t))|\mathbf{X}((k-1)\Delta t) = \mathbf{x}_i]$ can be calculated analytically. The calculation is detailed as follows.

$$
\begin{aligned}
&\mathrm{E}[G_j(\mathbf{X}((k+1)\Delta t))|\mathbf{X}(k\Delta t) = \mathbf{x}_i] \\
&= \int_{\mathbb{R}^n} G(\mathbf{x}, \boldsymbol{\mu}_j, \boldsymbol{\Sigma}_j)p(\mathbf{x}, (k+1)\Delta t|\mathbf{x}_i, k\Delta t)d\mathbf{x} \\
&= \int_{\mathbb{R}^n} G(\mathbf{x}, \boldsymbol{\mu}_j, \boldsymbol{\Sigma}_j)G(\mathbf{x}, \mathbf{A}_i, \mathbf{B}_i)d\mathbf{x} \\
&= \frac{1}{\sqrt{(2\pi)^{2n}\det(\boldsymbol{\Sigma}_j + \mathbf{B}_j)}} \\
&\quad \times \exp\left[-\frac{1}{2}(\mathbf{A}_i - \boldsymbol{\mu}_j)^T(\boldsymbol{\Sigma}_j + \mathbf{B}_i)^{-1}(\mathbf{A}_i - \boldsymbol{\mu}_j)\right] \\
&= G(\mathbf{x}_i, \boldsymbol{\mu}_j - \mathbf{a}(\mathbf{x}_i)\Delta t, \boldsymbol{\Sigma}_j + \boldsymbol{\sigma}\boldsymbol{\sigma}^T(\mathbf{x}_i)\Delta t).
\end{aligned}
\tag{9.23}
$$

The analytical expression for the function $e_j(\mathbf{x}_i)$ is obtained as,

$$
e_j(\mathbf{x}_i) = e^{-\bar{c}\Delta t}G(\mathbf{x}_i, \boldsymbol{\mu}_j - \mathbf{a}(\mathbf{x}_i)\Delta t, \boldsymbol{\Sigma}_j + \boldsymbol{\sigma}\boldsymbol{\sigma}^T(\mathbf{x}_i)\Delta t). \tag{9.24}
$$

Therefore, there is no longer a need for simulations to compute the expectation in the loss function (9.20) from the Feynman-Kac formula.

Let $\mathbf{w}(k) = [w_j(k)] \in \mathbb{R}^{N_G\times 1}$, $\mathbf{p}(k) = [p(\mathbf{x}_i, k\Delta t)] \in \mathbb{R}^{N_s\times 1}$, $\mathbf{G} = [G_{ij}] = [G_i(\mathbf{x}_j)] \in \mathbb{R}^{N_G\times N_s}$ and $\mathbf{E} = [e_{ij}] = [e_i(\mathbf{x}_j)] \in \mathbb{R}^{N_G\times N_s}$, we can rewrite the loss function (9.20) together with the normalization condition (9.16) in the matrix form as,

$$
\begin{aligned}
J(\mathbf{w}(k), \lambda(k)) &= \frac{1}{2}(\mathbf{p}(k-1) - \mathbf{E}^T\mathbf{w}(k))^T(\mathbf{p}(k-1) - \mathbf{E}^T\mathbf{w}(k)) \\
&\quad + \lambda(k)\left(\sum_{j=1}^{N_G} w_j(k) - 1\right) \\
&= \frac{1}{2}\big[\mathbf{w}^T(k)\mathbf{E}\mathbf{E}^T\mathbf{w}(k) - 2\mathbf{w}^T(k)\mathbf{E}\mathbf{p}(k-1) \\
&\quad + \mathbf{p}^T(k-1)\mathbf{p}(k-1)\big] \\
&\quad + \lambda(k)\left(\sum_{j=1}^{N_G} w_j(k) - 1\right), \quad k \geq 1.
\end{aligned} \tag{9.25}
$$

The optimal weight coefficients are determined to satisfy the necessary conditions for minimization of the loss:

$$
\begin{aligned}
\frac{\partial J(\mathbf{w}(k), \lambda(k))}{\partial \mathbf{w}(k)} &= \mathbf{E}\mathbf{E}^T\mathbf{w}(k) - \mathbf{E}\mathbf{p}(k-1) + \lambda(k)\mathbf{e}_w = 0, \\
\frac{\partial J(\mathbf{w}(k), \lambda(k))}{\partial \lambda(k)} &= \sum_{j=1}^{N_G} w_j(k) - 1 = 0,
\end{aligned} \tag{9.26}
$$

where $\mathbf{e}_w \in \mathbb{R}^{N_G \times 1}$ is a column vector of ones with the same size as $\mathbf{w}$.

We should note that the optimal weight coefficients can also be obtained by making use of the gradient search algorithms.

To extend the Feynman-Kac formula to the steady-state, we assume that $T \to \infty$ and $k \to \infty$ while Δt is kept finite. We further assume that the stationary solution of the PDF exists. Let $p_{ss}(\mathbf{x})$ denote the stationary PDF. Then, Eq. (9.13) reads,

$$
p_{ss}(\mathbf{x}) = \mathrm{E}\left[e^{-\int_0^{\Delta t} c(\mathbf{X}(s))ds} p_{ss}(\mathbf{X})\right]. \tag{9.27}
$$

Another way to reach the steady-state for autonomous or periodic systems is to consider the solution over consecutive finite time intervals $[k\Delta t, (k+1)\Delta t]$ $(k = 0, 1, 2, \ldots)$. This approach would lead to mappings of $p(\mathbf{x}, k\Delta t) \to p(\mathbf{x}, (k+1)\Delta t)$. Such a mapping described in the discretized state space has been known as the generalized cell mapping (GCM) for stochastic systems [31]. In the framework of cell mapping, the stationary response can be obtained with the help of one step transient probability matrix, which we shall discuss later.

9.3.4 Computation of Optimal Weights

Let D_G denote the domain where the system stays with probability not less than 99%. We discretize the domain into uniform grids, which represent the means of Gaussian neurons. The standard deviations for all the Gaussian functions are set equal to the grid size.

The optimal weight coefficients $\mathbf{w}(k)$ can be computed in Algorithm 1.

Algorithm 1 The RBFNN method based on the Feynman-Kac formula for the FPK equation.

1: **procedure** THE FPK EQUATION $\mathcal{L}_{FPK}[p(\mathbf{x}, t|\mathbf{x}_0, t_0)]$, THE NUMBER OF NEURONS N_G, THE NUMBER OF SAMPLING POINTS N_s, THE NUMBER OF INTERVALS N_t, THE TIME STEP SIZE Δt AND THE INITIAL PDF $\mathbf{p}(0)$.
2: The transient PDF $\mathbf{p}(k)$ at different time steps.
3: Discretize the domain D_G with N_G grid points. Each grid point will have one Gaussian neuron.
4: Uniformly sample N_s states $\mathbf{x}_i$ in a domain D_s containing D_G.
5: Calculate $\mathbf{G} = [G_{ij}] = [G_i(\mathbf{x}_j)]$ in Eq. (9.15).
6: Compute $\mathbf{E} = [e_{ij}] = [e_i(\mathbf{x}_j)]$ in Eq. (9.24).
7: **for** $k = 1$ to N_t **do**
8: Consider time intervals $[k\Delta t, (k+1)\Delta t]$.
9: Calculate the optimal weight coefficients $\mathbf{w}(k)$ using Eq. (9.26).
10: Use $\mathbf{p}(k) = \mathbf{G}^T\mathbf{w}(k)$ as the initial condition for the next time step.

9.4 Equation-Based RBFNN Method

Before presenting the computational examples, we briefly review the method of directly solving the FPK equation using RBFNN, referred to as the equation-based RBFNN method. Let the RBFNN in Eq. (9.14) be the trial solution of the FPK equation at the kth time step. Consider a finite difference scheme to approximate the time derivative as follows,

$$\frac{\partial \bar{p}}{\partial t} = \frac{1}{\Delta t}[\bar{p}(\mathbf{x}, \mathbf{w}(k)) - p(\mathbf{x}, (k-1)\Delta t)]. \tag{9.28}$$

By substituting the RBFNN trial solution shown in Eq. (9.14) and the time derivative shown in Eq. (9.28) into the FPK Eq. (9.4), we obtain a local residual $r(\mathbf{x}, \mathbf{w}(k))$ as a function of the weight parameters $\mathbf{w}(k)$,

$$r(\mathbf{x}, \mathbf{w}(k)) = p((k-1)\Delta t, \mathbf{x}) - \sum_{j=1}^{N_G} w_j(k) s_j(\mathbf{x}), \tag{9.29}$$

$$s_j(\mathbf{x}) = -\mathcal{L}_{FPK}[G(\mathbf{x}, \boldsymbol{\mu}_j, \boldsymbol{\Sigma}_j)]\Delta t + G(\mathbf{x}, \boldsymbol{\mu}_j, \boldsymbol{\Sigma}_j). \tag{9.30}$$

We define the loss function as

$$\begin{aligned} J(\mathbf{w}(k), \lambda(k)) &= \frac{1}{2}\sum_{j=1}^{N_s} r^2(\mathbf{x}_j, \mathbf{w}(k)) + \lambda(k)\left(\sum_{i=1}^{N_G} w_i(k) - 1\right) \\ &= \frac{1}{2}\big[\mathbf{w}^T(k)\mathbf{S}\mathbf{S}^T\mathbf{w}(k) - 2\mathbf{w}^T(k)\mathbf{S}\mathbf{p}(k-1) \\ &+ \mathbf{p}^T(k-1)\mathbf{p}(k-1)\big] \\ &+ \lambda(k)\left(\sum_{j=1}^{N_G} w_j(k) - 1\right), \quad k \geq 1, \end{aligned} \tag{9.31}$$

where $\mathbf{S} = [s_{ij}] = [s_i(\mathbf{x}_j)] \in \mathbb{R}^{N_G \times N_s}$. Although both the methods employ the same RBFNN, they represent fundamentally distinct computational paradigms.

The effectiveness and accuracy of the equation-based RBFNN method have been well-documented in previous studies [24–27]. Therefore, the equation-based RBFNN method will serve as a substitute for MCS to evaluate the proposed RBFNN method based on the Feynman-Kac formula.

9.5 Error Metric

We apply the proposed method to obtain steady-state and transient solutions of the FPK equation for three distinct systems without directly using the FPK equation. These systems include a strongly nonlinear system, a vibro-impact non-smooth system and a high-dimensional system.

To evaluate the accuracy of the proposed method, we define two errors. The first is the root mean square (RMS) error of the FPK equation,

$$J_{FPK} = \sqrt{\int_{\mathbb{R}^n} \left| \frac{\partial \bar{p}_{\Delta t}(\mathbf{x}, t)}{\partial t} - \mathcal{L}_{FPK}\big(\bar{p}_{\Delta t}(\mathbf{x}, t)\big) \right|^2 d\mathbf{x}}, \tag{9.32}$$

where $\bar{p}_{\Delta t}(\mathbf{x}, t)$ represents the PDF derived from the proposed method with the time step size Δt.

The second error metric, denoted as J_{PDF}, represents the discrepancy between the PDFs derived from the Feynman-Kac formula and those obtained by the equation-based RBFNN method.

$$J_{PDF} = \sqrt{\int_{\mathbb{R}^n} |p(\mathbf{x}, t) - p^*(\mathbf{x}, t)|^2 d\mathbf{x}}, \tag{9.33}$$

where $p^*(\mathbf{x}, t)$ denotes the PDF derived via the equation-based RBFNN method. These errors are also valid for steady-state cases when $t \to \infty$ and $\frac{\partial \bar{p}_{\Delta t}(\mathbf{x},t)}{\partial t} = 0$.

9.6 Examples of Feynman-Kac

9.6.1 Van der Pol System

We first consider a strongly nonlinear Van der Pol system subject to both additive and multiplicative Gaussian noises.

$$\begin{aligned} \frac{dX_1}{dt} &= X_2, \\ \frac{dX_2}{dt} &= -\beta(X_1^2 - 1)X_2 - X_1 \\ &\quad + X_1 W_1(t) + X_2 W_2(t) + W_3(t), \end{aligned} \tag{9.34}$$

where $W_i(t)$ represent independent Gaussian white noises with intensities $2D_i$. The drift and diffusion terms of the FPK equation of the Van der Pol system are shown as follows,

$$\begin{aligned} m_1 &= x_2, \\ m_2 &= -\beta(x_1^2 - 1)x_2 - x_1 + D_2 x_2, \\ b_{22} &= 2D_1 x_1^2 + 2D_2 x_2^2 + 2D_3. \end{aligned} \tag{9.35}$$

The parameters are set as $\beta = 1$ and $2D_i = 0.2$. The domain for Gaussian neurons is $D_G = [-3.2, 3.2] \times [-6.4, 6.4]$ divided into $N_G = 61 \times 61$ grids, where N_G represents the total number of Gaussian neurons. The domain for sampling points is $D_S = [-4, 4] \times [-8, 8]$ with $N_s = 81 \times 81$ points uniformly distributed in the domain. For the sake of fair comparison, the equation-based RBFNN method uses the same parameters and settings.

The computation of expectation of the Feynman-Kac formula is over a time step Δt. The integration of the function $c(\mathbf{X}(t))$ and short-time Gaussian approximation both introduce an error of order Δt. A proper choice of Δt should balance the solution accuracy and computational cost. We shall examine the effect of Δt in the numerical studies.

9.6.1.1 Stationary Response

We first report the stationary response of the Van der Pol system. Figure 9.1 presents the PDFs obtained using the proposed RBFNN method based on the Feynman-

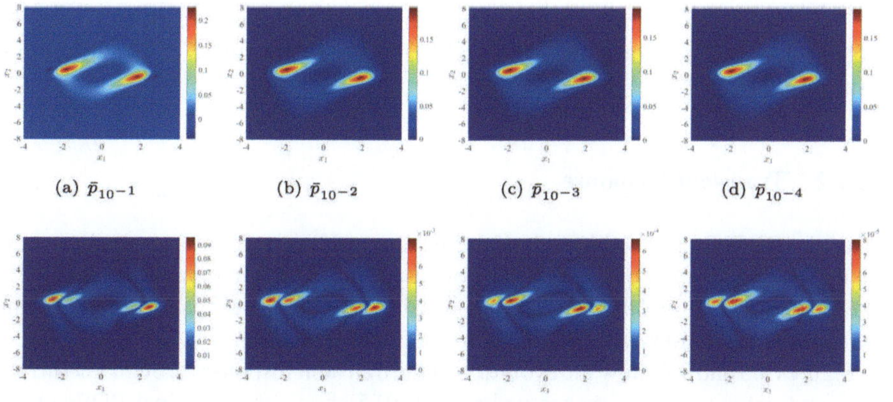

(a) $\bar{p}_{10^{-1}}$ (b) $\bar{p}_{10^{-2}}$ (c) $\bar{p}_{10^{-3}}$ (d) $\bar{p}_{10^{-4}}$

(e) $\|\bar{p}_{10^{-1}} - p^*\|$ (f) $\|\bar{p}_{10^{-2}} - p^*\|$ (g) $\|\bar{p}_{10^{-3}} - p^*\|$ (h) $\|\bar{p}_{10^{-4}} - p^*\|$

Fig. 9.1 The PDFs obtained by the proposed method and the differences to the equation-based RBFNN result for the Van der Pol system in Eq. (9.34). **a–d** correspond to time steps $\Delta t = [10^{-1}, 10^{-2}, 10^{-3}, 10^{-4}]$. **e–f** show the differences between the results obtained by proposed method and the equation-based RBFNN result

Fig. 9.2 Variation of errors with time step Δt. Red and blue curves denote the errors defined in Eqs. (9.32) and (9.33). Black dashed line represents the FPK error of the solution obtained by the equation-based RBFNN method

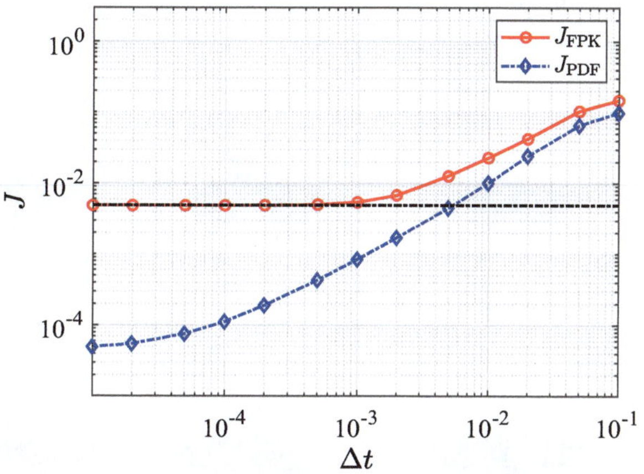

Kac formula for various time steps $\Delta t = [10^{-1}, 10^{-2}, 10^{-3}, 10^{-4}]$, along with their discrepancies to the PDFs derived from the equation-based RBFNN method. It is noticeable that the difference decreases with the reduction in time step. Moreover, when $\Delta t \leq 10^{-3}$, the results of the proposed method exhibit good agreement with the equation-based RBFNN result.

Figure 9.2 shows the effect of the time step Δt on the errors defined earlier. It appears that $\Delta t = 10^{-3}$ is a turning point of J_{FPK} and can be considered as optimal in terms of the balance of accuracy and efficiency. This is also confirmed by the results in Fig. 9.1.

The total computational times for calculating the stationary responses with the proposed method and the equation-based RBFNN method are 1.7687 and 2.3383 s, respectively.

9.6.1.2 Transient Response

Next, the RBFNN method based on the Feynman-Kac formula is applied to study the transient responses of the Van der Pol system. The time interval is set to [0, 10]. The time step is $\Delta t = 10^{-3}$. We take a Gaussian PDF with the mean $\boldsymbol{\mu}_0 = [0, 0]$ and the standard deviation $\boldsymbol{\sigma}_0 = [0.1, 0.1]$ as the initial condition.

Figures 9.3 and 9.4 show the evolution of the transient response through 3D surface plots and color contours, respectively. Figure 9.5 shows the marginal PDFs of the transient response of the Van der Pol system.

These results clearly demonstrate that the proposed method with a time step 10^{-3} yields results that are in close agreement with those obtained from the equation-based RBFNN method. The total computational times for calculating the transient responses with the proposed method and the equation-based RBFNN method are 2393 and 2386 s, respectively.

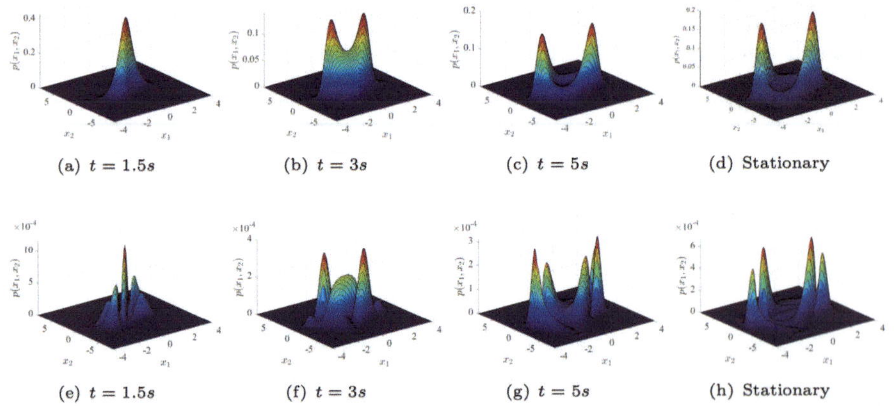

Fig. 9.3 3D surface plots of PDFs and differences to the equation-based RBFNN result of the Van der Pol system. **a–d** are the transient PDFs at $t = 1.5s$, $3s$, $5s$ and the stationary PDF, respectively, obtained by the proposed method. **e–h** show the differences between the results obtained by proposed method and the equation-based RBFNN results. The peaks of the error are of order 10^{-4}

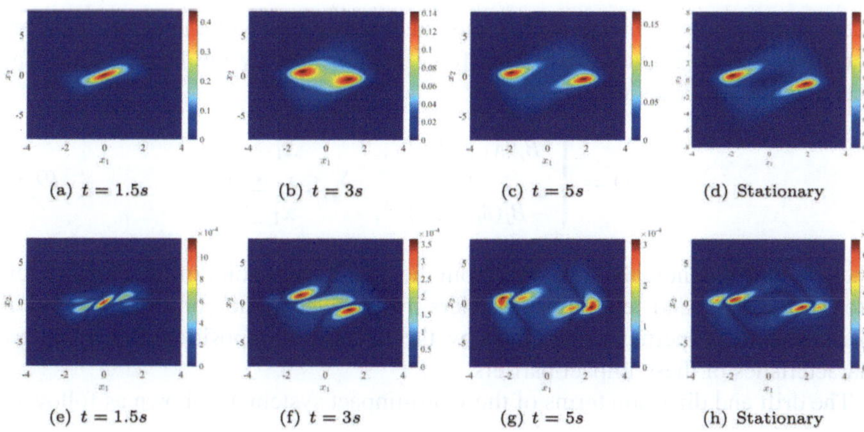

Fig. 9.4 Color contours of PDFs and differences to the equation-based RBFNN result of the Van der Pol system. **a–d** are the transient PDFs at $t = 1.5s$, $3s$, $5s$ and the stationary PDF, respectively, obtained by the proposed method. **e–h** show the differences between the results obtained by proposed method and the equation-based RBFNN results

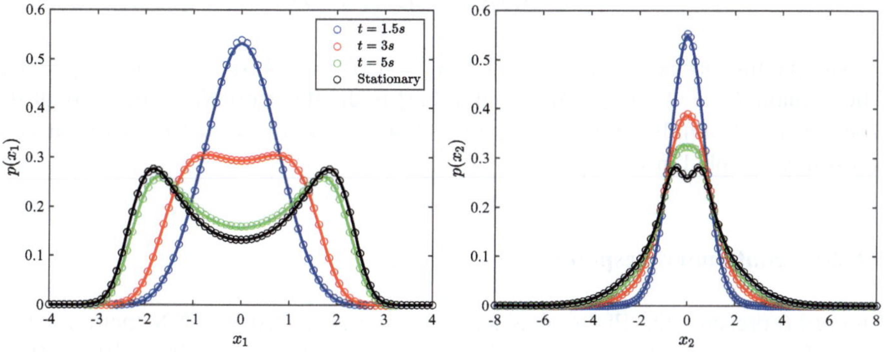

Fig. 9.5 The x_1 and x_2 marginal PDFs of the Van der Pol system at different times. **a** $p_{x_1}(\mathbf{x}_1)$. **b** $p_{x_2}(\mathbf{x}_2)$. Lines: By the proposed method. Circles: By the equation-based RBFNN method

9.6.2 Nonlinear Vibro-Impact System

Next, we consider a tri-stable vibro-impact system subject to both additive and multiplicative Gaussian noises.

$$\begin{aligned}\frac{dX_1}{dt} &= X_2, \\ \frac{dX_2}{dt} &= -\beta_1 X_2 - f(X_1) - \alpha_1 X_1 - \alpha_3 X_1^3 \\ &\quad - \alpha_5 X_1^5 + X_1 W_1(t) + X_2 W_2(t) + W_3(t),\end{aligned} \tag{9.36}$$

where $W_i(t)$ represent independent Gaussian white noises with intensities $2D_i$, and $f(x_1)$ denotes the impact force. This force is described by the Hertz contact law when the mass collides with the barrier.

$$f(x_1) = \begin{cases} B_r(x_1 - \delta_r)^{1.5}, & x_1 \geq \delta_r \\ 0, & \delta_l \leq x_1 \leq \delta_r \\ -B_l(\delta_l - x_1)^{1.5}, & x_1 \leq \delta_l \end{cases} \tag{9.37}$$

where δ_r and δ_l denote the distances from the equilibrium point of the impact oscillator to the right and left impact barriers, respectively. The constants B_r and B_l correspond to properties determined by the material composition and geometric characteristics of these impact barriers.

The drift and diffusion terms of the vibro-impact system are shown as follows,

$$\begin{aligned} m_1 &= x_2, \\ m_2 &= -\beta_1 x_2 - f(x_1) - \alpha_1 x_1 \\ &\quad - \alpha_3 x_1^3 - \alpha_5 x_1^5 + D_2 x_2, \\ b_{22} &= 2D_1 x_1^2 + 2D_2 x_2^2 + 2D_3. \end{aligned} \tag{9.38}$$

The parameters are set as $\beta_1 = 0.2$, $\alpha_1 = 1$, $\alpha_3 = -4$, $\alpha_5 = 1$ and $2D_i = 0.2$. The domain $D_G = [-3.2, 3.2] \times [-6.4, 6.4]$ is divided into $N_G = 61 \times 61$ grids. The domain of sampling points is $D_S = [-4, 4] \times [-8, 8]$. $N_s = 81 \times 81$ points are uniformly sampled in D_s.

9.6.2.1 Stationary Response

Figure 9.6 presents the PDFs obtained using the proposed RBFNN method based on the Feynman-Kac formula for various time steps $\Delta t = [10^{-1}, 10^{-2}, 10^{-3}, 10^{-4}]$. These results clearly demonstrate that for systems with multiple equilibrium states, the errors introduced by a larger Δt can lead to incorrect predictions, causing the steady-state response to erroneously converge to incorrect equilibrium states. This highlights the importance of selecting an appropriate Δt for accurate prediction in such complex systems.

The variation of errors with Δt is illustrated in Fig. 9.7. The result indicates again that $\Delta \leq 10^{-3}$ is optimal with a good balance of accuracy and efficiency.

For the proposed method and the equation-based RBFNN method, the total computational times for calculating the stationary responses are 1.3746 and 5.2292 s, respectively.

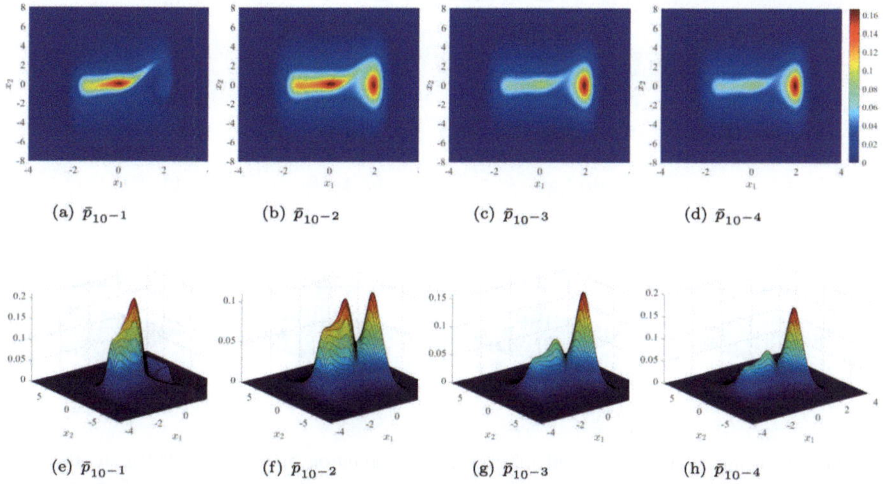

Fig. 9.6 The stationary PDF solutions of the vibro-impact system using the proposed method for various time steps $\Delta t = [10^{-1}, 10^{-2}, 10^{-3}, 10^{-4}]$. **a–d** are color contours. **e–f** are 3D surface plots

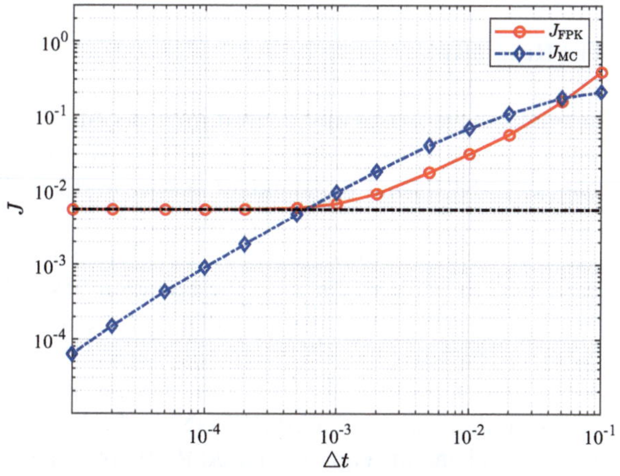

Fig. 9.7 Variation of the error with time step Δt for the Vibro-impact system. Red and blue curves denote the errors defined in Eqs. (9.32) and (9.33). Black dashed line represents the FPK error of the accurate solution derived by the equation-based RBFNN method

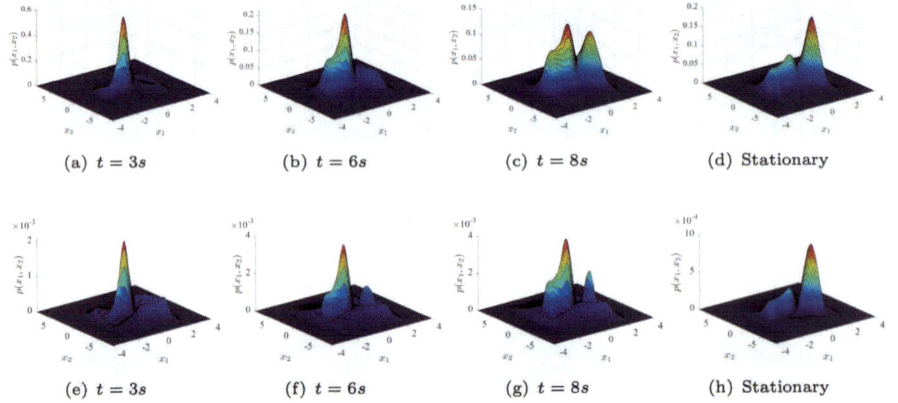

Fig. 9.8 3D surface plots of PDFs and differences to the equation-based RBFNN result for the vibro-impact system. **a–d** are the transient PDFs at $t = 3s, 6s, 8s$ and the stationary PDF, respectively, obtained by the proposed method. **e–h** show the differences between the results obtained by proposed method and the equation-based RBFNN results. The peaks of the error are of order 10^{-3}

9.6.2.2 Transient Response

Next, we study the transient responses of the vibro-impact system. The time interval is set as [0, 10] and the time step is chosen as $\Delta t = 10^{-3}$. We consider a Gaussian PDF with the mean $\boldsymbol{\mu}_0 = [0, 0]$ and the standard deviation $\boldsymbol{\sigma}_0 = [0.1, 0.1]$ as the initial condition.

Figures 9.8 and 9.9 illustrate the transient response evolution using the 3D surface plots and color contours, in comparison with the equation-based RBFNN result. Figure 9.10 shows the evolution of the marginal PDFs of the transient response of the vibro-impact system. These results clearly demonstrate that the proposed method, when employed with a time step size of 10^{-3} yields results that are in close agreement with those obtained from the equation-based RBFNN method.

For the proposed method and the equation-based RBFNN method, the total computational times for calculating the transient responses are 2561 and 2570 s, respectively.

9.6.3 Two Coupled Duffing Oscillators

As the last example, we consider a 4D coupled-Duffing system to demonstrate the effectiveness of the proposed method in high-dimensional systems. The governing equations are given by,

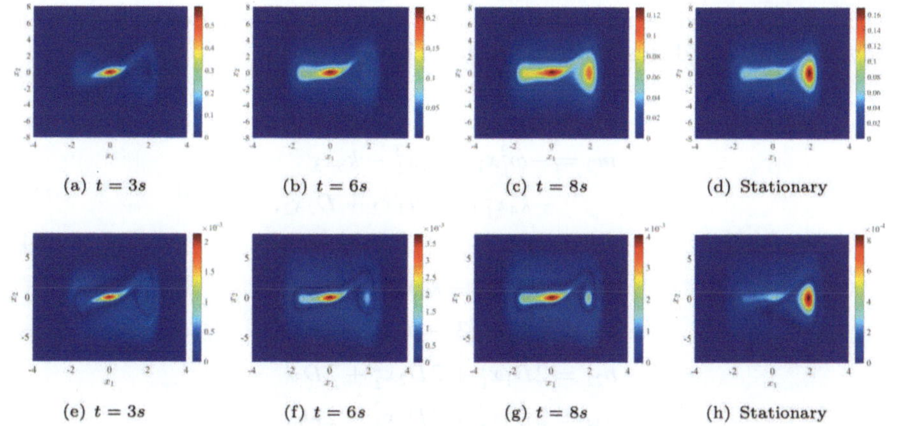

Fig. 9.9 Color contours of PDFs and differences to the equation-based RBFNN result for the vibro-impact system. **a–d** are the transient PDFs at $t = 3s, 6s, 8s$ and the stationary PDF, respectively, obtained by the proposed method. **e–h** show the differences between the results obtained by proposed method and the equation-based RBFNN results

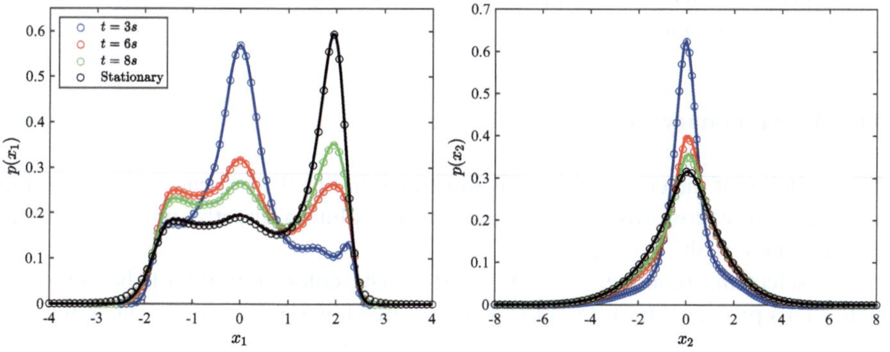

Fig. 9.10 The marginal PDFs of the Vibro-impact system for x_1 and x_2 at different instants. **a** The marginal PDF $p_{x_1}(\mathbf{x}_1)$. **b** The marginal PDF $p_{x_2}(\mathbf{x}_2)$. Lines: solutions obtained by the proposed method. Circles: solutions obtained by the equation-based RBFNN method

$$
\begin{aligned}
\frac{dX_1}{dt} &= X_2, \\
\frac{dX_2}{dt} &= -\omega_1^2 X_1 - k_1 X_1^3 - k_3 X_3 - k_4 X_1^2 X_3 \\
&\quad - \mu_1 X_2 + X_1 W_1(t) + X_2 W_2(t) + W_3(t), \\
\frac{dX_3}{dt} &= X_4, \\
\frac{dX_4}{dt} &= -\omega_2^2 X_3 - k_2 X_3^3 - k_3 X_1 - k_4 X_1^3/3 \\
&\quad - \mu_2 X_4 + X_3 W_4(t) + X_4 W_5(t) + W_6(t),
\end{aligned}
\tag{9.39}
$$

where $W_i(t)$ are independent and zero-mean Gaussian white noises with intensities $2D_i$. The drift and diffusion terms are given by

$$\begin{aligned}
m_1 &= x_2,\\
m_2 &= -\omega_1^2 x_1 - k_1 x_1^3 - k_3 x_3\\
&\quad - k_4 x_1^2 x_3 - \mu_1 x_2 + D_2 x_2,\\
m_3 &= x_4,\\
m_4 &= -\omega_2^2 x_3 - k_2 x_3^3 - k_3 x_1\\
&\quad - k_4 x_1^3/3 - \mu_2 x_4 + D_2 x_2,\\
b_{22} &= 2D_1 x_1^2 + 2D_2 x_2^2 + 2D_3\\
b_{44} &= 2D_4 x_3^2 + 2D_5 x_4^2 + 2D_6.
\end{aligned} \tag{9.40}$$

The parameters are $k_1 = 0.3, k_2 = 0.5, k_3 = 0.3, k_4 = 0.12, \mu_1 = 0.2, \mu_2 = 0.2, \omega_1 = 0.2, \omega_2 = 0.4$ and $2D_i = 0.04$. The domain $D_G = [-2, 2]^4$ is divided into $N_G = 16^4$ grids. The sampling domain is $D_S = [-2.5, 2.5]^4$. $N_s = 21^4$ points are uniformly sampled in D_s. The equation-based RBFNN method with the same settings is utilized for comparison.

9.6.3.1 Stationary Response

The variation of the errors with Δt is illustrated in Fig. 9.11. It can be observed that J_{FPK} converges towards the RMS error of the solution directly obtained from the FPK equation with $\Delta \leq 10^{-3}$.

We select the time step size $\Delta t = 10^{-3}$. The color contours of the stationary joint PDFs projected to different sub-spaces are shown in Fig. 9.12. The agreement

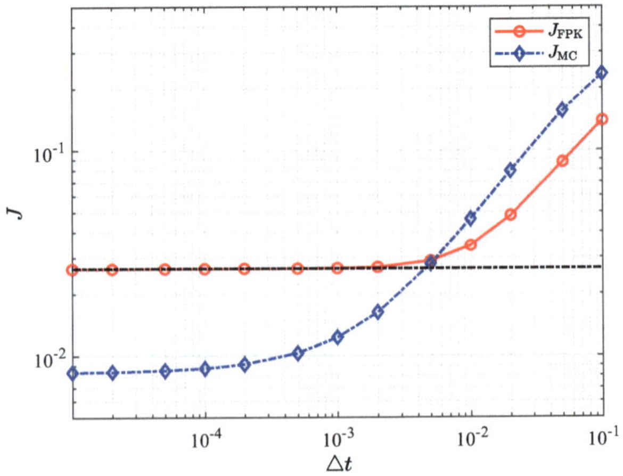

Fig. 9.11 Variation of errors with Δt for the 4D coupled Duffing system. Red and blue curves denote the errors defined in Eqs. (9.32) to (9.33). Black dashed line represents the FPK error of the accurate solution by the equation-based RBFNN method

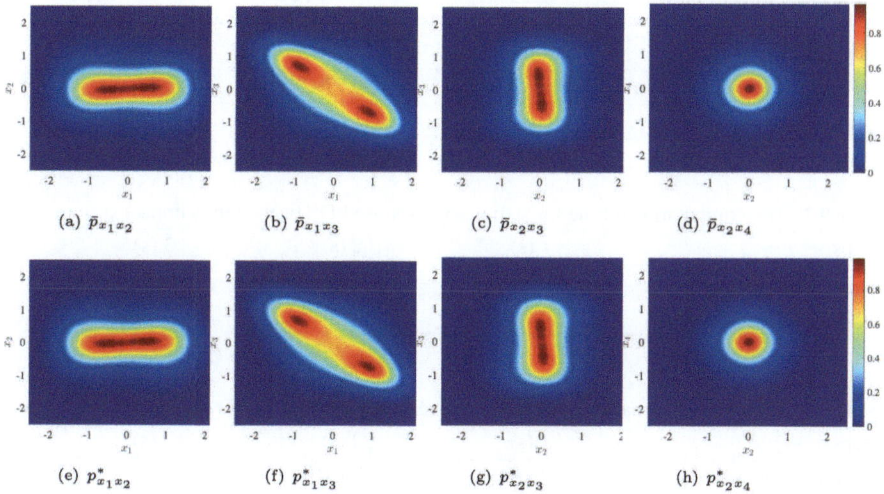

Fig. 9.12 Color contours of the stationary joint PDFs projected to $x_1 - x_2$, $x_1 - x_3$, $x_2 - x_3$ and $x_2 - x_4$ subspaces. **a–d** display the joint PDFs by the proposed method. **e–h** display the joint PDFs by the equation-based RBFNN method

between the proposed method and the equation-based RBFNN results is very good. The total computational times for calculating the stationary responses with the proposed method and the equation-based RBFNN method are 2671 and 3592 s, respectively.

We skip the transient responses of this example for the sake of length of the paper.

9.7 Comparison of Computational Time

Finally, we analyze the computational time for both the proposed method and the equation-based RBFNN method. We focus on the results for stationary response PDFs. It can be observed that the main time-consuming steps of both the methods can be divided into three parts:

1. Compute the matrix **G**.
2. Compute the matrix **E**.
3. Search for the optimal coefficients.

The computational times required for the stationary responses of three examples are listed in Tables 9.1, 9.2 and 9.3.

Since both the methods make use of RBFNN, the computational times for the first and third steps are similar. However, it is observable that the Feynman-Kac method significantly reduces computational time in the second step compared to the equation-based RBFNN method. This reduction is attributed to the fact that the Feynman-Kac

Table 9.1 The computational time for stationary response PDF of the Van der Pol system

Steps	1 (s)	2 (s)	3 (s)
Feynman-Kac method	0.6268	0.7792	0.2393
Equation-based RBFNN	0.6067	1.3299	0.2488

Table 9.2 The computational time for stationary response PDF of the vibro-impact system

Steps	1 (s)	2 (s)	3 (s)
Feynman-Kac method	0.5575	0.5834	0.2337
Equation-based RBFNN	0.5116	4.4613	0.2463

Table 9.3 The computational time for stationary response PDF of the 4D coupled duffing system

Steps	1 (s)	2 (s)	3 (s)
Feynman-Kac method	26.6453	54.1013	2547
Equation-based RBFNN	25.8577	1071	2581

method does not involve derivative calculations in computing the expected function shown in Eq. (9.19). These examples offer compelling evidence of efficiency of the Feynman-Kac method.

9.8 A Summary

The numerical method based on the Feynman-Kac formula is capable of solving parabolic PDEs without directly dealing with the equation. This is made possible by the martingale property of the process $\mathcal{M}(t)$, which provides an implicit integral equation describing the evolution of the PDF starting from a given initial condition. The neural networks can be used to approximate the general and unknown PDF of the stochastic process satisfying the integral equation. The STGA technique finds a new application in the machine learning era to deliver analytical expressions for the expectation of the stochastic martingale process $\mathcal{M}(t)$, thus eliminating the need for numerical simulations.

Since the integral equation is equivalent to the FPK equation, the solution of the FPK equation through the integral equation is obtained without using the FPK equation. The challenging examples of nonlinear stochastic systems demonstrate well the Feynman-Kac method. The results are compared with the equation-based RBFNN solution in terms of accuracy and efficiency. It is found that the Feynman-Kac method is quite competitive and promising for analyzing high-dimensional nonlinear stochastic systems. Many theoretical issues of the Feynman-Kac method are yet to be addressed including stability and convergence of the method.

References

1. Karatzas I, Shreve S (1991) Brownian motion and stochastic calculus, vol 113. Springer Science and Business Media, Berlin
2. Pham H (2009) Continuous-time stochastic control and optimization with financial applications, vol 61. Springer Science and Business Media, Berlin
3. Pardoux E, Peng SG (1990) Adapted solution of a backward stochastic differential equation. Syst Control Lett 14(1):55–61. https://doi.org/10.1016/0167-6911(90)90082-6
4. Pardoux E, Peng SG (1992) Backward stochastic differential equations and quasilinear parabolic partial differential equations. In: Rozovskii BL, Sowers RB (eds) Stochastic partial differential equations and their applications. Springer, Berlin, pp 200–217
5. Xu Y (2015) A complex Feynman-Kac formula via linear backward stochastic differential equations. arXiv:1505.03590
6. Alghassi H, Deshmukh A, Ibrahim N, Robles N, Woerner S, Zoufal C (2022) A variational quantum algorithm for the Feynman-Kac formula. Quantum 6:730
7. Pham H (2015) Feynman-Kac representation of fully nonlinear PDEs and applications. Acta Math Vietnam 40:255–269
8. Nguwi JY, Penent G, Privault N (2023) A fully nonlinear Feynman-Kac formula with derivatives of arbitrary orders. J Evol Equ 23(1):22
9. Bouchard B, Touzi N (2004) Discrete-time approximation and Monte-Carlo simulation of backward stochastic differential equations. Stoch Process Their Appl 111(2):175–206. https://doi.org/10.1016/j.spa.2004.01.001
10. Hwang CO, Mascagni M, Given JA (2003) A Feynman-Kac path-integral implementation for Poisson's equation using an h-conditioned Green's function. Math Comput Simul 62(3):347–355. https://doi.org/10.1016/S0378-4754(02)00224-0
11. Dalang RC, M ller C, Tribe R (2008) A Feynman-Kac-type formula for the deterministic and stochastic wave equations and other P.D.E.'s. https://doi.org/10.1090/S0002-9947-08-04351-1
12. Richter L, Berner J (2022) Robust SDE-based variational formulations for solving linear PDEs via deep learning. In: Proceedings of the 39th international conference on machine learning, PMLR 162, Baltimore, Maryland, USA, pp 1–16
13. Park KMS, Stinchcombe AR (2022) Deep reinforcement learning of viscous incompressible flow. J Comput Phys 467(111):455. https://doi.org/10.1016/j.jcp.2022.111455
14. Han J, Nica M, Stinchcombe AR (2020) A derivative-free method for solving elliptic partial differential equations with deep neural networks. J Comput Phys 419(109):672. https://doi.org/10.1016/j.jcp.2020.109672
15. Han J, Jentzen A, E W (2018) Solving high-dimensional partial differential equations using deep learning. PNAS 115(34):8505–8510
16. Berg J, Nyström K (2018) A unified deep artificial neural network approach to partial differential equations in complex geometries. Neurocomputing 317:28–41. https://doi.org/10.1016/j.neucom.2018.06.056
17. Chen X, Yang L, Duan J, Karniadakis GE (2021) Solving inverse stochastic problems from discrete particle observations using the Fokker-Planck equation and physics-informed neural networks. SIAM J Sci Comput 43(3):B811–B830. https://doi.org/10.1137/20M1360153
18. Zhang H, Xu Y, Li Y, Kurths J (2020) Statistical solution to SDEs with α-stable Lévy noise via deep neural network. Int J Dyn Control 8(4):1129–1140. https://doi.org/10.1007/s40435-020-00677-0
19. Xu Y, Zhang H, Li Y, Zhou K, Liu Q, Kurths J (2020) Solving Fokker-Planck equation using deep learning. Chaos: Interdiscip J Nonlinear Sci 30(1). https://doi.org/10.1063/1.5132840
20. Zhang H, Xu Y, Liu Q, Wang X, Li Y (2022) Solving Fokker-Planck equations using deep KD-tree with a small amount of data. Nonlinear Dyn 108(4):4029–4043. https://doi.org/10.1007/s11071-022-07361-2

21. Zhang H, Xu Y, Liu Q, Li Y (2023) Deep learning framework for solving Fokker-Planck equations with low-rank separation representation. Eng Appl Artif Intell 121(106):036. https://doi.org/10.1016/j.engappai.2023.106036
22. Sirignano J, Spiliopoulos K (2018) DGM: a deep learning algorithm for solving partial differential equations. arXiv:1708.07469v5 [q-fin.MF]
23. Hu Z, Zhang J, Ge Y (2021) Handling vanishing gradient problem using artificial derivative. IEEE Access 9:22,371–22,377. https://doi.org/10.1109/ACCESS.2021.3054915
24. Wang X, Jiang J, Hong L, Sun JQ (2022) Random vibration analysis with radial basis function neural networks. Int J Dyn Control 10(5):1385–1394. https://doi.org/10.1007/s40435-021-00893-2
25. Wang X, Jiang J, Hong L, Zhao A, Sun JQ (2023) Radial basis function neural networks solution for stationary probability density function of nonlinear stochastic systems. Probab Eng Mech 71(103):408. https://doi.org/10.1016/j.probengmech.2022.103408
26. Qian J, Chen L, Sun JQ (2023) Transient response prediction of randomly excited vibro-impact systems via RBF neural networks. J Sound Vib 546(117):456. https://doi.org/10.1016/j.jsv.2022.117456
27. Wang X, Jiang J, Hong L, Sun JQ (2022) First-passage problem in random vibrations with radial basis function neural networks. J Vib Acoust 144(5):051,014. https://doi.org/10.1115/1.4054437
28. Sun JQ, Hsu CS (1990) The generalized cell mapping method in nonlinear random vibration based upon short-time Gaussian approximation. J Appl Mech 57(4):1018–1025
29. Park J, Sandberg IW (1991) Universal approximation using radial-basis-function networks. Neural Comput 3:246–257
30. Risken H (1984) The Fokker-Planck equation - methods of solution and applications. Springer, New York
31. Sun JQ (2006) Stochastic dynamics and control. Elsevier Science Ltd., Oxford, U.K

Chapter 10
Green's Operator

The method of Feynman-Kac formula discussed in Chap. 9 translates the process of obtaining the solution of PDEs, e.g. the FPK equation, to the problem of solving an integral equation without the need to evaluate derivatives for the PDEs. The resulting integral equation involves the expectation of nonlinear functions and the unknown probability density function. The numerical solution of the integral equation is still a non-trivial task.

This chapter introduces another method to convert general partial differential equations to integral equations with the help of Green's function. The method, motivated by the work on neural operator in the literature, is called Green's operator. We apply the method to the FPK equation and provide examples to demonstrate the method. In particular, we express the unknown PDF solution of the FPK equation in terms of a RBFNN. The solution of the integral equation is obtained by training the neural network to minimize the equation error.

The RBFNN method discussed in Chap. 8 is applied directly to solve for the FPK equation in various applications. The method is an example of so-called physics-informed neural networks (PINNs) [1–4]. PINNs have been demonstrated to be a highly viable tool for analyzing scientific and engineering problems including fluid mechanics [5–7], material science [8], and biology [9, 10].

PINNs face challenges related to scalability, efficiency, and generalization, when applied to deal with problems of varying boundary conditions, initial conditions, or system parameter change. Deep neural networks are usually used to attack these issues, which can be computationally costly, particularly for complex or multi-query tasks [11–13]. Neural operators are developed to address these issues [14].

We make use the idea of neural operator and the property of Green's functions of the boundary value problems of PDEs to convert the problem of obtaining solutions of PDEs to that of integral equations.

J.-Q. Sun, *Random Vibration with Machine Learning Method*,
Nonlinear Physical Science, https://doi.org/10.1007/978-981-95-0812-9_10

10.1 Boundary Value Problem

The boundary value problem in many engineering application is governed by PDEs subject to boundary and initial conditions. Since PDEs are classified into different types, we consider a PDE in a general operator form. Consider a general PDE for a function $p(\mathbf{x}, t)$ defined on a d-dimensional spatial domain $\mathbf{x} \in \Omega \subseteq \mathbb{R}^d$ and over a finite time interval $t \in [0, T]$.

$$\begin{aligned} \mathcal{M}[p(\mathbf{x}, t)] &= \mathcal{L}[p(\mathbf{x}, t)] + \mathcal{N}[p(\mathbf{x}, t)] \\ &\quad + u(\mathbf{x}, t), \ \ \mathbf{x} \in \Omega, \ \ t \in [0, T] \quad (10.1) \\ \mathcal{B}_j[p(\mathbf{x}, t)] &= b_j(\mathbf{x}, t), \ \ \mathbf{x} \in \partial\Omega, \ \ j = 1, 2, \cdots \quad (10.2) \\ p(\mathbf{x}, t_0) &= p(\mathbf{x}_0), \quad (10.3) \end{aligned}$$

where $\mathcal{M}$ is an inertial operator, $\mathcal{L}$ is a linear spatial operator, $\mathcal{N}$ is a nonlinear operator, and $u(\mathbf{x}, t)$ is a prescribed external input. $\mathcal{B}_j$ is an operator defined the boundary $\partial\Omega$ of the domain. $b_j(\mathbf{x}, t)$ is a non-homogeneous boundary term. The number of boundary conditions and the order of boundary operators can be specified for a given PDE. We consider one simple initial condition as an example, which implies that the inertial operator $\mathcal{M}$ is first order in time. This is always true if $p(\mathbf{x}, t_0)$ is treated as a state vector of functions.

Define a linear operator for the PDE as

$$\mathcal{P}[p(\mathbf{x}, t)] = \mathcal{M}[p(\mathbf{x}, t)] - \mathcal{L}[p(\mathbf{x}, t)]. \quad (10.4)$$

Consider the solution of the following problem, known as the Green's function

$$\begin{aligned} \mathcal{P}[G(\mathbf{x} - \mathbf{x}_0, t - t_0)] &= \delta(\mathbf{x} - \mathbf{x}_0)\delta(t - t_0), \quad (10.5) \\ &\quad \mathbf{x}, \mathbf{x}_0 \in \Omega, \ \ t, t_0 \in [0, T] \\ \mathcal{B}_j[G(\mathbf{x} - \mathbf{x}_0, t - t_0)] &= 0, \ \ \mathbf{x} \in \partial\Omega, \ \ j = 1, 2, \ldots \quad (10.6) \\ G(\mathbf{x} - \mathbf{x}_0, 0) &= 0, \ \ \mathbf{x} \in \partial\Omega. \quad (10.7) \end{aligned}$$

Let $p_h(\mathbf{x}, t)$ denote the solution satisfying all boundary and initial conditions such that,

$$\begin{aligned} \mathcal{P}[p_h(\mathbf{x}, t)] &= \mathbf{e}_h(\mathbf{x}, t), \quad (10.8) \\ &\quad \mathbf{x}_0 \in \Omega, \ \ t, t_0 \in [0, T] \\ \mathcal{B}_j[p_h(\mathbf{x}, t)] &= b_j(\mathbf{x}, t), \ \ \mathbf{x} \in \partial\Omega, \ \ j = 1, 2, \cdots \quad (10.9) \\ p_h(\mathbf{x}, t_0) &= p(\mathbf{x}_0), \quad (10.10) \end{aligned}$$

where $\mathbf{e}_h(\mathbf{x}, t)$ is the error of the homogeneous equation. $\mathbf{e}_h(\mathbf{x}, t) \neq 0$ when $p_h(\mathbf{x}, t)$ satisfies all the boundary conditions only and is not a solution of the homogeneous equation. When $p_h(\mathbf{x}, t)$ is the exact solution of the homogeneous equation, the equation error is zero $\mathbf{e}_h(\mathbf{x}, t) = 0$.

Assume that a closed-form solution of the problem defined by Eqs. (10.5)–(10.7), or a highly accurate numerical solution, for example, in terms of neural networks, can be found. The solution of Eqs. (10.1), (10.2) and (10.3) can be written as

$$p(\mathbf{x}, t) = p_h(\mathbf{x}, t) + \int_0^t \int_\Omega G(\mathbf{x} - \mathbf{x}_0, t - t_0) s(\mathbf{x}_0, t_0) d\mathbf{x}_0 dt_0, \tag{10.11}$$

where we have defined a symbol as,

$$s(\mathbf{x}, t) = \mathcal{N}[p(\mathbf{x}, t)] + u(\mathbf{x}, t) - \mathbf{e}_h(\mathbf{x}, t). \tag{10.12}$$

By way of construction, the solution $p(\mathbf{x}, t)$ in Eq. (10.11) satisfies all the boundary and initial conditions. Applying the operator $\mathcal{P}[\cdot]$ to Eq. (10.11), we can show that Eq. (10.1) is satisfied. The PDE is now converted to an integral equation.

10.2 Algebraic Solution for Linear Problem

The price to pay for moving away from the PDE is that the integral equation automatically requires to compute the global solution in the domain. Assume that the computational domain of interest is a finite subset $\mathbb{D} \subset \Omega$.

$$p(\mathbf{x}, t) = p_h(\mathbf{x}, t) + \int_0^t \int_\mathbb{D} G(\mathbf{x} - \mathbf{x}_0, t - t_0) s(\mathbf{x}_0, t_0) d\mathbf{x}_0 dt_0. \tag{10.13}$$

Since the integration with respect to $\mathbf{x}_0$ is in the domain $\mathbb{D}$, the solution $p(\mathbf{x}_0, t_0)$ must be available for all $\mathbf{x}_0 \in \mathbb{D}$ and $0 \leq t_0 \leq t$. Let $\mathbf{x}_i$ $(1 \leq i \leq N_t)$ be a sample or grid point in $\mathbb{D}$ for numerical integration where N_t is the total number of sample points. The solution $p(\mathbf{x}, t)$ must be computed at all sample points $\mathbf{x}_i \in \mathbb{D}$ at time t. When the dimension of the space Ω is high, the computational effort to solve the integral equation can be excessive.

We first focus on linear PDEs and point out an interesting and important feature of the integral equation. Equation (10.13) provides a time domain integration for the solution.

Consider the solution of $p(\mathbf{x}, k\Delta t)$, where Δt is a small time step, and $k \geq 0$ is an integer. For $k \geq 1$, we have

$$\begin{aligned}p(\mathbf{x}, k\Delta t) &= p_h(\mathbf{x}, k\Delta t) + \int_0^{k\Delta t}\int_{\mathbb{D}} G(\mathbf{x}-\mathbf{x}_0, t-t_0)s(\mathbf{x}_0, t_0)d\mathbf{x}_0 dt_0 \\ &= p_h(\mathbf{x}, k\Delta t) \\ &+ \Delta t \int_{\mathbb{D}} \Big[\frac{1}{2}G(\mathbf{x}-\mathbf{x}_0, k\Delta t)s(\mathbf{x}_0, 0) \\ &+ G(\mathbf{x}-\mathbf{x}_0, (k-1)\Delta t)s(\mathbf{x}_0, \Delta t) + \cdots \\ &+ G(\mathbf{x}-\mathbf{x}_0, \Delta t)s(\mathbf{x}_0, (k-1)\Delta t) \\ &+ \frac{1}{2}G(\mathbf{x}-\mathbf{x}_0, 0)s(\mathbf{x}_0, k\Delta t)\Big]d\mathbf{x}_0. \end{aligned} \tag{10.14}$$

It is noted that this iterative solution as time domain integration using Simpson's rule is accurate to the order Δt^2. This is one order of magnitude better than the path integral solution [15]. Equation (10.14) holds for general cases when the Green's operator and the source term $s(\mathbf{x}, t)$ may not be in a separable form and are time-varying. For example, the FPK equation can be for a system subject to both white noise random excitation and periodic forcing. In this case, $s(\mathbf{x}, t)$ is time-varying via the dependence on $p(\mathbf{x}, t)$ and due to the periodic functions in drift and diffusion terms.

When the unknown solution $p(\mathbf{x}, t)$ appears in all the operators linearly, the above iterative equations contain linear terms of $p(\mathbf{x}, k\Delta t)$. Hence, after properly collecting the terms and introducing the sample points $\mathbf{x}_i \in \mathbb{D}$, $p(\mathbf{x}_i, k\Delta t)$ can be computed from all the algebraic equation without the need for further numerical iterations.

Recall that the term $s(\mathbf{x}, k\Delta t)$ and the solution $p(\mathbf{x}, k\Delta t)$ must be sampled at the same set of points $\mathbf{x}_i \in \mathbb{D}$. This leads to a linear algebraic equation for the nodal values $p(\mathbf{x}_i, k\Delta t)$ of the solution.

The FPK equation for random vibration problem is linear in the PDF of the response. Hence, even for highly nonlinear dynamic systems, the response PDF can always be computed with the above computational procedure by solving a linear algebraic equation with a truncation error in time domain of order Δt^2. This is indeed a remarkable simplification and advancement in obtaining transient PDF solutions of nonlinear stochastic dynamic systems.

10.3 Neural Network Solution for Non-linear Problem

For the system where the solution $p(\mathbf{x}, t)$ appears in the equation nonlinearly, e.g. $p^2(\mathbf{x}, t)$, Eq. (10.11) is implicit in terms of the unknown solution $p(\mathbf{x}, t)$. To obtain numerical solution from this equation, we consider neural networks.

Let $p(\mathbf{x}, t)$ be expressed as a neural network with two inputs $[\mathbf{x}, t]$ and one output p. Let all trainable coefficients be denoted as $\mathbf{w}$, and the neural network solution be denoted as $p_{NN}(\mathbf{x}, t, \mathbf{w})$. Define the error of Eq. (10.11) at a sample point $[\mathbf{x}_i, t_i] \in$

$\Omega \times [0, T]$,

$$\begin{aligned} e_i(\mathbf{x}_i, t_i, \mathbf{w}) &= p_{NN}(\mathbf{x}_i, t_i, \mathbf{w}) - p_h(\mathbf{x}_i, t_i) \\ &- \int_0^{t_i}\int_\Omega G(\mathbf{x}_i, t_i, \mathbf{x}_0, t_0)\Big[\mathcal{N}[p_{NN}(\mathbf{x}_0, t_0, \mathbf{w})] \\ &+ u(\mathbf{x}_0, t_0) - \mathbf{e}_h(\mathbf{x}_0, t_0)\Big]d\mathbf{x}_0 dt_0, \end{aligned} \tag{10.15}$$

where $1 \leq i \leq N_s$ and N_s is the total number of sample points. When the domain $\Omega \times [0, T]$ is predefined, it can be discretized so that the sampling points are well distributed in the domain and the integration in the equation can be numerically evaluated.

A loss function for the neural network solution can be defined as

$$J(\mathbf{w}) = \sum_{i=1}^{N_s} e_i^2(\mathbf{x}_i, t_i, \mathbf{w}). \tag{10.16}$$

Any mature training algorithm can now be applied to find the optimal coefficients $\mathbf{w}$ to minimize the loss.

Recall that $p(\mathbf{x}, t)$ and therefore $p_{NN}(\mathbf{x}, t, \mathbf{w})$ automatically satisfy all the boundary and initial conditions in Eqs. (10.2) and (10.3) by way of construction of the homogeneous solution $p_h(\mathbf{x}, t)$. It is highlighted again that because of the Green's operator formulation that eliminates the differential equation and reduces the order of differentiations in the solution process, and the introduction of the particular solution $p_h(\mathbf{x}, t)$ that satisfies all the initial and boundary conditions, only one loss function is needed without the consideration of weighting different loss factors. As shall be demonstrated later, simplicity of loss function often leads to efficiency of computation.

In order to achieve computational efficiency, it is also important to be able to obtain a closed-form analytical solution of the Green's operator. When the differential operator $\mathcal{P}[\cdot]$ does not render a closed-form solution of the Green's operator, we can modify the original differential equation to create a new operator that does provide a closed-form solution. We shall illustrate how to do this in the examples.

10.4 Examples of Green's Operator

10.4.1 Linear Random Vibration Problem

The equation of motion for a linear oscillator subject to white noise excitation can be written in the state-space form as,

$$dX_1 = X_2\,dt, \qquad (10.17)$$
$$dX_2 = (-2\zeta\omega_0 X_2 - \omega_0^2 X_1)\,dt + d\beta(t),$$

where $\beta(t)$ is a Wiener process such that

$$\mathrm{E}[d\beta(t)] = 0, \;\; \mathrm{E}[d\beta(t)d\beta(s)] = \begin{cases} 2D\,dt, & t = s, \\ 0, & t \neq s. \end{cases} \qquad (10.18)$$

The formal derivative of $\beta(t)$ is known as white noise. Let $\mathbf{X} = [X_1, X_2]^T$ be the state vector of the system. In the matrix form, the state equation can be expressed as the stochastic equation in the Itô sense.

$$d\mathbf{X} = \mathbf{A}\mathbf{X}\,dt + \mathbf{B}\,d\beta(t), \qquad (10.19)$$

where

$$\mathbf{A} = \begin{bmatrix} 0 & 1 \\ -\omega_0^2 & -2\zeta\omega_0 \end{bmatrix}, \;\; \mathbf{B} = \begin{bmatrix} 0 \\ 1 \end{bmatrix}. \qquad (10.20)$$

The associated FPK equation for the response PDF reads

$$\begin{aligned} \frac{\partial}{\partial t} p_{\mathbf{X}}(\mathbf{x}, t|\mathbf{x}_0, t_0) &= -\frac{\partial}{\partial x_1}[x_2 p_{\mathbf{X}}] \\ &\quad - \frac{\partial}{\partial x_2}[(-2\zeta\omega_0 x_2 - \omega_0^2 x_1) p_{\mathbf{X}}] + D\frac{\partial^2 p_{\mathbf{X}}}{\partial x_2^2}, \end{aligned} \qquad (10.21)$$

where $\mathbf{x} = [x_1, x_2]^T$. Rearranging terms, we have

$$\begin{aligned} \frac{\partial p_{\mathbf{X}}}{\partial t} - D\frac{\partial^2 p_{\mathbf{X}}}{\partial x_2^2} &= -\frac{\partial}{\partial x_1}[x_2 p_{\mathbf{X}}] \\ &\quad - \frac{\partial}{\partial x_2}[(-2\zeta\omega_0 x_2 - \omega_0^2 x_1) p_{\mathbf{X}}] \\ &\equiv -\nabla \cdot [\mathbf{A}\mathbf{x} p_{\mathbf{X}}]. \end{aligned} \qquad (10.22)$$

The left-hand side of the equation defines the PDE operator which involves the diffusion in x_2 direction only. The Green's function can be found as,

$$G(\mathbf{x} - \mathbf{x}_0, t - t_0) = \frac{\delta(x_1 - x_{10})}{\sqrt{4\pi D(t - t_0)}} \exp\left(-\frac{(x_2 - x_{20})^2}{4D(t - t_0)}\right). \qquad (10.23)$$

Because the operator on the right hand side of Eq. (10.22) does not have diffusion in x_1 direction, it is *degenerate*. Without including the diffusion in x_1 direction, the Green's function in Eq. (10.23) is found to be ineffective to capture the diffusion in x_1 direction and offers a PDF solution with poor accuracy.

There are two ways to address the issue degenerate operator. We shall introduce both below.

10.4.1.1 ε Diffusion Approach

We can introduce a small diffusion term in x_1 direction as

$$\frac{\partial p_{\mathbf{X}}}{\partial t} - \left(\varepsilon \frac{\partial^2 p_{\mathbf{X}}}{\partial x_1^2} + D \frac{\partial^2 p_{\mathbf{X}}}{\partial x_2^2}\right) = -\nabla \cdot [\mathbf{A}\mathbf{x} p_{\mathbf{X}}], \tag{10.24}$$

where $0 < \varepsilon \ll 1$. The Green's function now becomes a full two-dimensional Gaussian function.

$$G_{\varepsilon}(\mathbf{x} - \mathbf{x}_0, t - t_0) = \frac{1}{4\pi\sqrt{D\varepsilon}(t - t_0)} \times \exp\left(-\frac{(x_1 - x_{10})^2}{4\varepsilon(t - t_0)} - \frac{(x_2 - x_{20})^2}{4D(t - t_0)}\right). \tag{10.25}$$

Using this Green's function, the solution of the FPK equation can be represented as a convolution integral.

$$p_{\mathbf{X}}(\mathbf{x}, t|\mathbf{x}_0, t_0) = \int_{t_0}^{t} \int_{\mathbb{R}^2} G_{\varepsilon}(\mathbf{x} - \mathbf{x}', t - t') \times \left(-\nabla' \cdot [\mathbf{A}\mathbf{x}' p_{\mathbf{X}}(\mathbf{x}', t'|\mathbf{x}_0, t_0)]\right) d\mathbf{x}' dt'. \tag{10.26}$$

The accuracy of the solution is much improved based on comparison with MCS results. The system parameters are taken as $\zeta = 0.1$, $\omega_0 = 1.0$, and $D = 0.1$ in the simulation.

Figure 10.1 presents a direct comparison between the original degenerate Green's function and the modified Green's function with different diffusion parameter ε. As can be seen from the figure, introduction of a small diffusion in x_1 direction retains the main feature of the Green's function and makes the solution more accurate as Fig. 10.2 shows that the PDFs at different time instances compare well with the simulation results.

For the proposed method, the spatial domain spans $[-2, 2]$ in both x_1 and x_2 directions, discretized into a 51×51 grid for spatial integration. The initial probability distribution is Gaussian with standard deviations $\sigma_1 = \sigma_2 = 0.2$. Time is discretized into $N_t = 300$ steps over a total duration $t = 30$ seconds, resulting in a time step $\Delta t = t/N_t$. The parameter ε is taken as $1e^{-3}$. The execution time is 52.93 s. The MCS results are generated using 10^7 simulated trajectories, utilizing the same spatial and temporal domains to estimate the probability distribution as those defined for

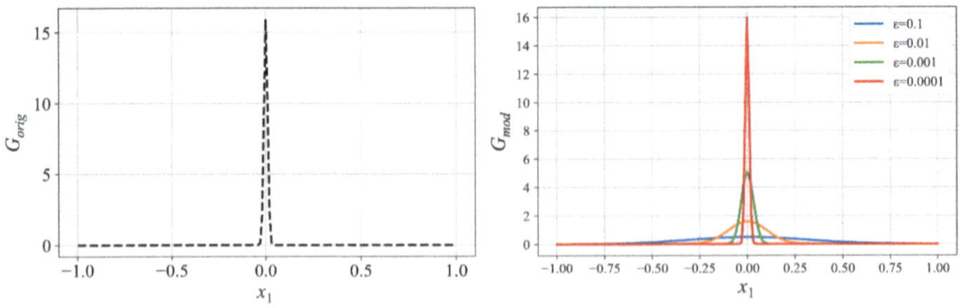

Fig. 10.1 Comparison of the original and modified Green's functions

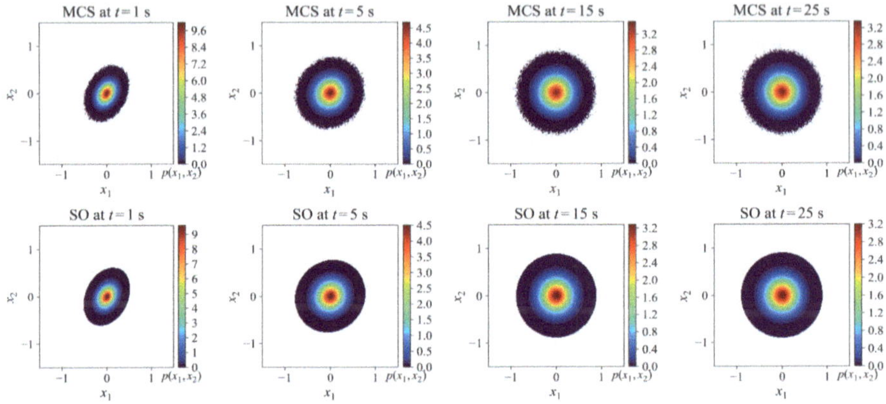

Fig. 10.2 Comparison of joint PDF $p_{\mathbf{X}}((x_1, x_2)$ of the system (10.17) obtained by MCS and the Green's operator with added diffusion at selected time instances. The first row shows the PDFs by MCS, while the second row displays the results by the Green's operator method

the proposed method. The execution time is 385.9 s. The proposed Green's operator approach is obviously efficient and accurate.

10.4.1.2 Modified Equation Approach

Another way to deal with degenerate operators is to patch the missing operator terms by modifying the original differential equation. For example, Eq. (10.22) can be written as

$$\frac{\partial p_{\mathbf{X}}}{\partial t} - D\left[\frac{\partial^2 p_{\mathbf{X}}}{\partial x_1^2} + \frac{\partial^2 p_{\mathbf{X}}}{\partial x_2^2}\right] = -\nabla \cdot [\mathbf{A}\mathbf{x}p_{\mathbf{X}}] - D\frac{\partial^2 p_{\mathbf{X}}}{\partial x_1^2}. \tag{10.27}$$

The Green's function is now a regular two-dimensional Gaussian function.

$$G(\mathbf{x}-\mathbf{x}_0, t-t_0) = \frac{1}{4\pi D(t-t_0)} \times \exp\left(-\frac{(x_1-x_{10})^2}{4D(t-t_0)} - \frac{(x_2-x_{20})^2}{4D(t-t_0)}\right). \tag{10.28}$$

The solution of the FPK equation in terms of the convolution integral reads.

$$p_{\mathbf{X}}(\mathbf{x}, t|\mathbf{x}_0, t_0) = \int_{t_0}^{t}\int_{\mathbb{R}^2} G(\mathbf{x}-\mathbf{x}', t-t') \times \Big(-\nabla'\cdot[\mathbf{A}\mathbf{x}' p_{\mathbf{X}}(\mathbf{x}', t'|\mathbf{x}_0, t_0)] - D\frac{\partial^2 p_{\mathbf{X}}(\mathbf{x}', t'|\mathbf{x}_0, t_0)}{\partial x_1^2}\Big) d\mathbf{x}' dt'. \tag{10.29}$$

The advantage of the modified equation approach lies in that it relaxes the requirement for grids to place the means of the Gaussian functions and makes the solution converges faster. The penalty is that more higher order derivatives of the solution are introduced in the modified equation, which has a potential to reduce the computational efficacy of the solution. The following table compares the two methods when applied to this example. The error term is the L^2 norm defined as

$$\epsilon_G = \left(\int_{\mathbb{R}^2} [p_{\mathbf{X}}(\mathbf{x}, t|\mathbf{x}_0, t_0) - p_{\text{MCS}}(\mathbf{x}, t|\mathbf{x}_0, t_0)]^2 \, d\mathbf{x}\right)^{1/2} \tag{10.30}$$

where $p_{\text{MCS}}(\mathbf{x}, t|\mathbf{x}_0, t_0)$ is the simulated PDF of $p_{\mathbf{X}}(\mathbf{x}, t|\mathbf{x}_0, t_0)$ obtained with the Green's operator approach (Table 10.1).

Table 10.1 Comparison of the CPU time and global error ϵ_G at $t = 30s$ of the solutions obtained with the Green's operator approach. The error is measured with respect to the Monte Carlo simulation result

Method	Time (s)	ϵ_G
MCS	137.8	0
ε Diffusion	15.61	1.2×10^{-4}
Modified equation	10.9	1.6×10^{-4}

10.4.2 Bouc-Wen System

The state-space equations for the Bouc-Wen system are

$$\begin{aligned}
\frac{dx}{dt} &= y, \\
\frac{dy}{dt} &= -2\zeta y - \alpha x - (1-\alpha)z + W(t), \\
\frac{dz}{dt} &= Ay - \beta z|y||z|^{n-1} - \gamma y|z|^n,
\end{aligned} \tag{10.31}$$

where $W(t)$ is Gaussian white noise with $E[W(t)W(s)] = 2D\delta(t-s)$. The corresponding FPK equation governing the probability density $p_{\mathbf{X}}(\mathbf{x}, t)$ is:

$$\frac{\partial p_{\mathbf{X}}}{\partial t} = -\nabla \cdot [\mathbf{f}(\mathbf{x}) p_{\mathbf{X}}] + D\frac{\partial^2 p_{\mathbf{X}}}{\partial y^2}, \tag{10.32}$$

where $\mathbf{x} = [x, y, z]^T$ and

$$\mathbf{f}(\mathbf{x}) = \begin{bmatrix} y \\ -2\zeta y - \alpha x - (1-\alpha)z \\ Ay - \beta z|y||z|^{n-1} - \gamma y|z|^n \end{bmatrix}.$$

The original FPK equation is degenerate due to the absence of diffusion in the x- and z-directions. We consider the ϵ diffusion method as follows.

$$\begin{aligned}
&\frac{\partial p_{\mathbf{X}}}{\partial t} - \left(\varepsilon_x \frac{\partial^2 p_{\mathbf{X}}}{\partial x^2} + D\frac{\partial^2 p_{\mathbf{X}}}{\partial y^2} + \varepsilon_z \frac{\partial^2 p_{\mathbf{X}}}{\partial z^2} \right) \\
&= -\nabla \cdot [\mathbf{f}(\mathbf{x}) p_{\mathbf{X}}],
\end{aligned} \tag{10.33}$$

where $\varepsilon_x, \varepsilon_z \ll 1$. This modification ensures a well-posed operator for constructing the Green's operator.

The modified Green's operator G_ε satisfies:

$$\begin{aligned}
&\left\{ \frac{\partial}{\partial t} - \varepsilon_x \frac{\partial^2}{\partial x^2} - D\frac{\partial^2}{\partial y^2} - \varepsilon_z \frac{\partial^2}{\partial z^2} \right\} G_\varepsilon \\
&= \delta(\mathbf{x} - \mathbf{x}')\delta(t - t'),
\end{aligned} \tag{10.34}$$

and its analytical solution is a three-dimensional Gaussian kernel:

$$G_\varepsilon(\mathbf{x}-\mathbf{x}', t-t') = \frac{1}{[4\pi\varepsilon_x D\varepsilon_z(t-t')]^{3/2}} \times \exp\left(-\sum_{i=1}^{3}\frac{(x_i-x_i')^2}{4\kappa_i(t-t')}\right), \tag{10.35}$$

where $\kappa_1 = \varepsilon_x$, $\kappa_2 = D$, $\kappa_3 = \varepsilon_z$. Using the modified Green's operator, the solution of the FPK equation is expressed as a convolution integral

$$p_{\mathbf{X}}(\mathbf{x}, t|\mathbf{x}_0, t_0) = \int_{t_0}^{t}\int_{\mathbb{R}^3} G_\varepsilon(\mathbf{x}-\mathbf{x}', t-t') \times \left(-\nabla'\cdot\left[\mathbf{f}(\mathbf{x}')p_{\mathbf{X}}(\mathbf{x}', t'|\mathbf{x}_0, t_0)\right]\right)d\mathbf{x}'dt'. \tag{10.36}$$

Selecting small parameters ε_x and ε_z requires balancing numerical stability with the need to preserve the near-degenerate behavior of the original model. A rigorous approach to estimate these parameters is through von Neumann stability analysis applied to the spatially discretized equation [16]. For the three-dimensional explicit scheme, each direction (x, y, z) contributes a dimensionless diffusion number defined as

$$\nu_\ell = \frac{\kappa_\ell \Delta t}{(\Delta x_\ell)^2}, \tag{10.37}$$

where κ_ℓ represents the diffusion coefficient in the ℓ-direction ($\kappa_x = \varepsilon_x$, $\kappa_y = D$, $\kappa_z = \varepsilon_z$), Δx_ℓ is the spatial mesh size, and Δt is the time step. To ensure numerical stability, the combined effect of diffusion across all directions must satisfy

$$\nu_x + \nu_y + \nu_z = \frac{\varepsilon_x \Delta t}{(\Delta x)^2} + \frac{D\Delta t}{(\Delta y)^2} + \frac{\varepsilon_z \Delta t}{(\Delta z)^2} \le c, \tag{10.38}$$

where the constant c is determined by the discretization scheme. For an explicit Euler method in three dimensions, von Neumann analysis dictates

$$c = \frac{1}{2}, \quad \nu_x \le \frac{1}{6}, \quad \nu_z \le \frac{1}{6}. \tag{10.39}$$

These constraints arise from the requirement that the amplification factor for all Fourier modes remains bounded. Violating these conditions by choosing excessively small ε_x or ε_z may destabilize the solution, while overly large values introduce unphysical smoothing that obscures the system's intrinsic dynamics.

To reconcile these competing demands, the artificial diffusion coefficients should be parameterized as

$$\varepsilon_x = \alpha_x \cdot \frac{(\Delta x)^2}{6\Delta t}, \quad \varepsilon_z = \alpha_z \cdot \frac{(\Delta z)^2}{6\Delta t}, \tag{10.40}$$

where $\alpha_x, \alpha_z \in (0,\ 1]$ are tuning factors. A conservative initial choice is

$$\alpha_x = \alpha_z = 0.5, \tag{10.41}$$

ensuring compliance with the stability bounds while minimizing artificial dissipation.

The Courant–Friedrichs–Lewy condition must be satisfied by the nonlinear drift function $\mathbf{f}(\mathbf{x})$ [16]

$$v \cdot \frac{\Delta t}{\min(\Delta x, \Delta y, \Delta z)} \leq 1, \tag{10.42}$$

where $v = \max(|\mathbf{f}(\mathbf{x})|)$. When v is large, Δt should be reduced with a higher priority than ε_x and ε_z.

The nonlinear hysteresis characteristics of the Bouc–Wen model are investigated with varying parameter combinations $(\beta + \gamma)$, and a focus on hardening/softening behaviors and topological evolution of restoring force probability density functions.

For the softening regime $(\beta + \gamma > 0)$, parameters are set as $n = A = 1, \beta = 0.5$, $\gamma = 0.5, \zeta = \alpha = 0.1, \tau = 0.01\,s$ with initial conditions defined by the Dirac function $p_{\mathbf{X}}(\mathbf{x}, t_0) = \delta(\mathbf{x} - \mathbf{x}_0)$ at $t_0 = 0$. Three excitation intensities are calculated over the predefined spatial domains $x \in [-3.0, 3.0]$, $y \in [-3.0, 3.0]$, $z \in [-2.0, 2.0]$ with grid resolutions $N = N_x \times N_y \times N_z = 40 \times 40 \times 30$. The parameters $\varepsilon_x = 0.00875$ and $\varepsilon_z = 0.00379$.

The marginal PDFs of displacement, velocity, and restoring force exhibit strong agreement with MCS results, as shown in Fig. 10.3.

The hardening regime $(\beta + \gamma < 0,\ \beta > \gamma)$ is analyzed by adjusting γ to -0.6 while keeping other parameters. The spatial domains are $x \in [-3.0, 3.0]$, $y \in [-3.0, 3.0]$, $z \in [-2.5, 2.5]$. The parameters $\varepsilon_x = 0.0197$ and $\varepsilon_z = 0.0248$. The marginal PDFs of displacement, velocity, and restoring force are illustrated in Fig. 10.4. The PDF of the restoring force in the hardened Bouc–Wen system displays a crisp unimodal distribution.

The Bouc-Wen system can also be studied with the modified equation approach. We list the key equations below.

$$\begin{aligned} &\frac{\partial p_{\mathbf{X}}}{\partial t} - D\left(\frac{\partial^2 p_{\mathbf{X}}}{\partial x^2} + \frac{\partial^2 p_{\mathbf{X}}}{\partial y^2} + \frac{\partial^2 p_{\mathbf{X}}}{\partial z^2}\right) \\ &= -\nabla \cdot [\mathbf{f}(\mathbf{x}) p_{\mathbf{X}}] - D\left(\frac{\partial^2 p_{\mathbf{X}}}{\partial x^2} + \frac{\partial^2 p_{\mathbf{X}}}{\partial z^2}\right). \end{aligned} \tag{10.43}$$

The Green's operator G of the modified FPK equation satisfies,

$$\begin{aligned} &\left\{\frac{\partial}{\partial t} - D\left[\frac{\partial^2}{\partial x^2} + \frac{\partial^2}{\partial y^2} + \frac{\partial^2}{\partial z^2}\right]\right\} G \\ &= \delta(\mathbf{x} - \mathbf{x}')\delta(t - t'), \end{aligned} \tag{10.44}$$

and its analytical solution is a three-dimensional Gaussian kernel:

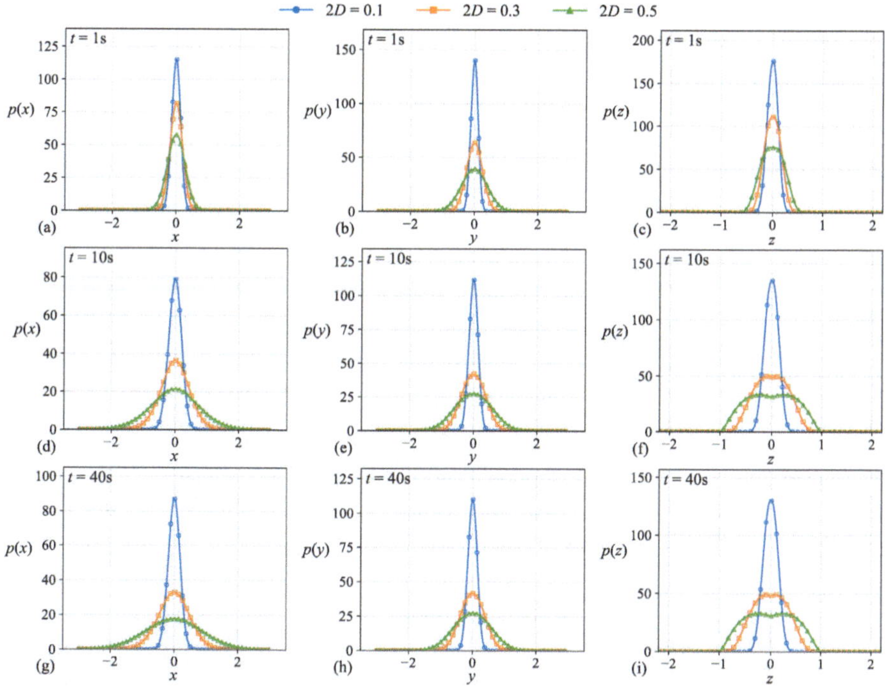

Fig. 10.3 The marginal probability density function of displacement, velocity and restoring force with $D = 0.05, 0.15, 0.25$ when $t = 1s, 10s, 40s$. Solid line: the proposed method results. Symbols: MC simulation results

$$G(\mathbf{x}-\mathbf{x}', t-t') = \frac{1}{[4\pi D(t-t')]^{3/2}} \times \exp\left(-\sum_{i=1}^{3}\frac{(x_i-x_i')^2}{4D(t-t')}\right), \tag{10.45}$$

where $x_1 = x$, $x_2 = y$ and $x_3 = z$. The solution of the FPK equation is expressed as a convolution integral

$$p_{\mathbf{X}}(\mathbf{x}, t|\mathbf{x}_0, t_0) = \int_{t_0}^{t}\int_{\mathbb{R}^3} G(\mathbf{x}-\mathbf{x}', t-t') \times\Big[-\nabla'\cdot\big[\mathbf{f}(\mathbf{x}')p_{\mathbf{X}}(\mathbf{x}', t'|\mathbf{x}_0, t_0)\big] - D\left(\frac{\partial^2 p_{\mathbf{X}}(\mathbf{x}', t'|\mathbf{x}_0, t_0)}{\partial x^2}+\frac{\partial^2 p_{\mathbf{X}}(\mathbf{x}', t'|\mathbf{x}_0, t_0)}{\partial z^2}\right)\Big]d\mathbf{x}'dt'. \tag{10.46}$$

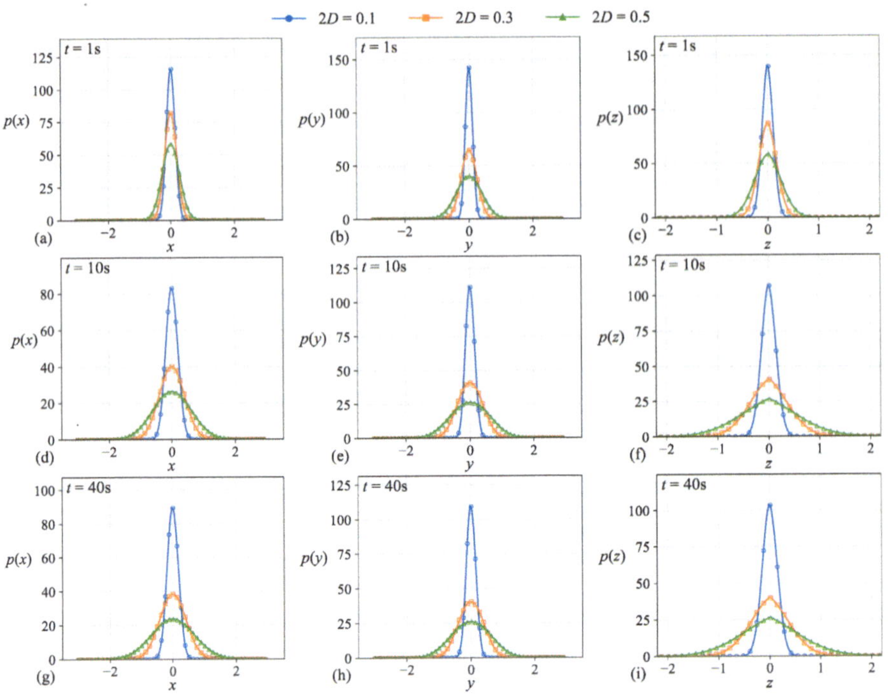

Fig. 10.4 The marginal probability density function of displacement, velocity and restoring force with $D = 0.05, 0.15, 0.25$ when $t = 1s, 10s, 40s$. Solid line: the proposed method results. Symbols: MC simulation results

Numerical experiments of this example also show that the modified equation approach is more efficient and accurate than the ε diffusion approach.

References

1. Raissi M, Perdikaris P, Karniadakis GE (2019) Physics-informed neural networks: a deep learning framework for solving forward and inverse problems involving nonlinear partial differential equations. J Comput Phys 378:686–707
2. Greenfeld D, Galun M, Basri R, Yavneh I, Kimmel R (2019) Learning to optimize multigrid PDE solvers. In: International conference on machine learning, PMLR, pp 2415–2423
3. Kochkov D, Smith JA, Alieva A, Wang Q, Brenner MP, Hoyer S (2021) Machine learning–accelerated computational fluid dynamics. Proc Natl Acad Sci 118(21):e2101784,118
4. Khoo Y, Lu J, Ying L (2021) Solving parametric PDE problems with artificial neural networks. Eur J Appl Math 32(3):421–435
5. Wu L, Kilingar NG, Noels L et al (2020) A recurrent neural network-accelerated multi-scale model for elasto-plastic heterogeneous materials subjected to random cyclic and non-proportional loading paths. Comput Methods Appl Mech Eng 369(113):234
6. Cai S, Mao Z, Wang Z, Yin M, Karniadakis GE (2021) Physics-informed neural networks (PINNs) for fluid mechanics: a review. Acta Mech Sin 37(12):1727–1738

7. Jagtap AD, Mao Z, Adams N, Karniadakis GE (2022) Physics-informed neural networks for inverse problems in supersonic flows. J Comput Phys 466(111):402
8. Chen Y, Lu L, Karniadakis GE, Dal Negro L (2020) Physics-informed neural networks for inverse problems in nano-optics and metamaterials. Opt Express 28(8):11,618–11,633
9. Daneker M, Zhang Z, Karniadakis GE, Lu L (2023) Systems biology: identifiability analysis and parameter identification via systems-biology-informed neural networks. In: Computational modeling of signaling networks. Springer, Berlin, pp 87–105
10. Daryakenari NA, De Florio M, Shukla K, Karniadakis GE (2024) Ai-aristotle: a physics-informed framework for systems biology gray-box identification. Plos Comput Biol 20(3)
11. Rao C, Sun H, Liu Y (2021) Physics-informed deep learning for computational elastodynamics without labeled data. J Eng Mech 147(8):04021,043
12. Haghighat E, Abouali S, Vaziri R (2023) Constitutive model characterization and discovery using physics-informed deep learning. Eng Appl Artif Intell 120(105):828
13. Li W, Bazant MZ, Zhu J (2021) A physics-guided neural network framework for elastic plates: comparison of governing equations-based and energy-based approaches. Comput Methods Appl Mech Eng 383:113,933
14. Lu L, Jin P, Karniadakis GE (2019) Deeponet: learning nonlinear operators for identifying differential equations based on the universal approximation theorem of operators. arXiv:1910.03193
15. Sun JQ (2006) Stochastic dynamics and control. Elsevier Science Ltd., Oxford, U.K.
16. LeVeque RJ (2007) Finite difference methods for ordinary and partial differential equations: steady-state and time-dependent problems. SIAM

Chapter 11
Data-Driven Random Vibration

This chapter discusses what we can do when we have a large assemble of sample responses of mechanical systems subject to random excitations. This is known as data-driven random vibration analysis. When the response process is reasonably assumed to be ergodic, we can record a very long sample response and use temporal average to approximate assemble mean, or break the long record to a large number of shorter and yet sufficient long trajectories to calculate assemble average. We should point out that it is not easy to collect a large set of responses of a system with uncertain parameters.

11.1 What Can We Do with Data?

Before we jump into the discussion of random vibration, let us briefly discuss what we can do with data from engineering systems.

11.1.1 System Property from Data

Data from well-designed sensors often contain rich information about the system. Let us take the time series as an example.

Spectral analysis of time series is a classical method to study measured data [1]. The spectral distribution of time series contains the knowledge about resonances, damping, degree of freedom of the system and response bandwidth. The asymptote of the spectral distribution at frequency $f \to 0$ and $f \to \infty$ has special meaning. The spectral level at frequency $f = 0$ is related to the static property of the system such as the displacement or stiffness. The slope of the asymptote as $f \to \infty$ in the log-log scale represents the order of the system or the relative degree. The slope can

J.-Q. Sun, *Random Vibration with Machine Learning Method*,
Nonlinear Physical Science, https://doi.org/10.1007/978-981-95-0812-9_11

also indicate whether the system is subject to broadband noise, is chaotic or fractional order.

Extensive response measurements of nonlinear systems can be used to conduct the global analysis in the state space [2, 3]. The global analysis discovers the attractors of the nonlinear system and the domains of attraction.

When the data are from a multiple input and multiple output system, the functional relationship among the data is often a focus of research. Correlation analysis of the data can reveal the cause-effect relationship between two measurements [1].

When designing controls for under-actuated systems, one can determine if an output is differentially flat or not, identify the relative degree of the system and construct a data-driven state feedback control by making use of the flat output [4, 5].

11.1.2 Physical Laws in Data

It is a common interest to identify the governing equation for the system behind the data. Engineering systems can be highly complex with time-varying nonlinearities and uncertainties. It is a challenge to develop accurate mathematical models of such systems. In the age of high-tech and super-computing, it has become increasingly feasible to use a large number of sensors to measure the input and output of complex systems over a long time. With the availability of the data, we should be able to identify the underlining mathematical model of the system and determine the model parameters without much knowledge about the system.

Data are abundant whereas models often remain elusive in some fields, as in climate science, neuroscience, ecology, finance, epidemiology, *etc.*. In biology, data-driven modeling of fish trajectories in group or individual stands for complex stochastic systems [6, 7]. Porfiri's group models an individual zebrafish trajectory through stochastic equations and Markov chain. The calibration of the model parameters through experimental data by the maximum likelihood and weighted least squares method allows studies of how the fish behaves in front of constrains [8]. In transportation, machine learning-based approaches aim at accurate trajectory prediction of aircraft to manage air traffic control [9] and of vehicles for safe and efficient autonomous driving [10].

Various system identification techniques have been developed over the years and are well documented in the survey papers [11, 12]. A comprehensive survey of nonlinear system identification techniques from 2006 to 2016 can be found in [13]. Using time-varying linear system model to describe the responses of nonlinear dynamic systems is frequently studied [14–16]. Factors including the looseness of structural joints, amplitude dependent material properties and boundary conditions with variable stiffness constraints make the behavior of engineering structures highly nonlinear and complex [17]. Other intrinsic nonlinear phenomena such as hysteresis, friction, inelasticity, harmonics, jumps or modal interactions call for nonlinear system model [13].

The system identification is often done in frequency-domain. A modified H_2 algorithm for frequency response function estimation is proposed in [18]. The modification of the algorithm introduces additional correlated output to improve the computational effectiveness. A generalized approach to apply the method of "Reverse Path" to continuous mechanical systems with multiple nonlinearities is introduced in [19].

The methods for linear systems are extended to nonlinear systems with the help of associated linear equations (ALEs) and associated frequency response functions (AFRFS) [20]. Some technical issues and limitations are addressed in this reference in order to apply the method to MDOF systems. Another technique in frequency domain interprets nonlinearities as feedback forces acting on the underlying linear system [21]. The system identification can also be done in time domain or in time-frequency mixed domain [22–24]. The method of nonlinear subspace identification is formulated in time and frequency domain for a strongly nonlinear satellite structure [25].

A general framework is still not available for nonlinear system identification problems. The well-known model of Nonlinear Auto-Regressive Moving Average with eXogenous Input (NARMAX) is considered for the detection of damage in engineering structures [26]. The work in [27] studies an amplitude and frequency modulation method (AFMM) for extracting characteristics of nonlinear systems and intermittent transient responses by processing stationary and transient responses using the empirical mode decomposition, Hilbert-Huang transform (HHT), and nonlinear dynamic characteristics derived from perturbation analysis. The experimental extraction of nonlinear normal modes (NNMs) under broadband forcing is investigated in [22], where a state-space model is established, and then is converted to the system in the modal space.

Nonparametric or black-box methods directly work with input-output data and are very popular. Some common examples include subspace identification, kernel methods, statistical learning, data mining, artificial neural networks (ANNs), auxiliary model identification, multi-innovation identification and hierarchical identification [28–33]. The black-box model does not need the prior knowledge of the system and sometimes has no clear physical meaning [17]. Raissi and colleagues have developed a deep learning based algorithm and illustrated the method on complex dynamic systems of Lorenz, fluid flow behind a cylinder, Hopf bifurcation and the Glycoltic oscillator [31].

A long short-term memory (LSTM) recurrent neural networks (RNNs) has been recently introduced to predict high-dimensional chaotic systems [34]. A reservoir-based computing algorithm gives impressive predictions for chaotic systems [35]. It should be noted that the high-dimensional neural network models are lack of physical meaning. As a tool for regression, neural networks also don't predict well out of the range of the training data set [17, 30]. The Bayesian framework has attracted a great attention from the research community [36–38]. This method tries to find the best fitting model among the proposed models by studying the occurrence probabilities of the proposed functions [39].

When the dynamic model of the system is represented by differential equations, we often have to create the time series of derivatives of the measured data. The algebraic method and operational calculus are excellent tools for computing derivatives of the measured signals by means of integration, thus to avoid amplifying the measurement noise in the data [40, 41]. The algebraic framework in [42] has been studied for identification of first-order linear systems. The algebraic identification has been used to estimate parameters of a servo motor [43], highly damped flexible structures [44], and mass-spring-damper systems [45]. In [46, 47], the algebraic method is applied to identify nonlinear dynamical systems.

Polynomials have been a popular functional choice for modeling nonlinear systems. From the point of view of fitting the data, whether we use polynomials, artificial neural networks or any other functions, as long as a sufficient number of functions are used, the resulting model will fit the data to any degree of accuracy.

In this case, the major concern is overfitting when the obtained model offers little flexibility to describe the dynamic behavior of the system that may not have been contained in the data used for the system modeling. For a given set of measurements, an important question to ask is: What is the simplest model that can fit the measured data sufficiently accurate and still provide enough flexibility to make reliable predictions with the unseen data?

Brunton and colleagues proposed a method called the sparse identification of nonlinear dynamics (SINDy) in 2016 for creating a sparse representation of the unknown nonlinear function of the system [48]. The method assumes that only a few important terms out of a library of functions are needed to describe the dynamics of the system. It combines machine learning and sparsity-promoting techniques to find the sparse representation in the space of possible functions to model nonlinear dynamical systems. The connection of the SINDy method to the Akaike information criteria (AIC) for model selection has been studied in [49]. The promising results for difficult system identification problems such as hybrid dynamical systems, chaotic Lorenz system and Burger's partial differential equation have been obtained [49, 50].

The SINDy method applied to the model predictive control delivers better performance, requires significantly less data, and is more computationally efficient and robust to noise than the neural networks model [51]. Studies of the SINDy method have reported the usage of high volume of time series data from different initial conditions for model selection and for cross-validation [49]. While this is feasible for simulation studies, it can be expensive for applications with real time data. Robustness to noise and requirement for measurements of derivatives are concerns with the SINDy approach [30]. The total variation regularized derivatives are commonly used to deal with these concerns [52, 53]. An integral form of equations of motion in combination with sparse regression is proposed in [54]. The SINDy approach is extended to combine the algebraic method and a K-fold cross validation approach to deal with measurement noise, initial conditions and differentiations of signals [46, 47].

From the material test data, the constitution law and symmetry of the material can be discovered.

11.1.3 Prediction Across Physical Domain

The most difficult and yet most meaningful data-driven research is to use data in one physical domain to predict what would happen in another physical domain. For example, one estimates fatigue and other damages from motion measurement. Another example is to identify and classify faults from the data of motion, sound and temperature.

In [55], [56], a detailed example of predicting high-speed train wheel wear from the motion data of the train is presented. It should be pointed out that the cross-domain prediction usually deals with very complex implicit functional relationship among different quantities such as motion and material wear. A hybrid approach is usually needed involving the use of extensive data and knowledge of mathematical model of the system.

In case of wheel wear prediction, contact mechanics and wear theory will be needed together with Newton's law or Lagrange dynamics and control theory. If neural networks are used to model, a deep network should be considered because of the complexity of the functional relationship between motion and wheel wear.

11.1.4 A Remark on Digital Twin

Digital twin of engineering system is a dynamic, time-varying and always-improving model when new data about the system become available. This is engineers' dream model of complex engineering system.

When a digital twin of a system involves only kinematic or motion prediction, the data needed to develop the digital twin is probably manageable.

However, for cross-domain prediction involving material property change such as fatigue damages, fracture and wear, there is not enough data from engineering system or laboratory to develop a neural network model for prediction.

It is expensive to measure various damages and their dependence on operation conditions and parameters. Some damages such as train wheel wear or system property changes such as fatigue build-up cannot be directly measured with today's sensing technology whether in the laboratory or in-situ. However, such so-called labeled data are crucial for developing and updating the data-driven neural networks model such as digital twin.

Another difficulty to develop a true digital twin of mechanical systems is due to a finite number of sensors that can be practically implemented. It is noted that real systems usually have infinite degrees of freedom. Hence, the data collected by the finite number of sensors may not discover all possible dynamics and damages of the system.

11.2 Response Moments from Data

Assume that we collect a large assemble of response trajectories $x_n(t_k)$ where $1 \leq n \leq N_s$ and $t_k = k\Delta t$ and Δt is the sample time. Te sample time $0 < \Delta t \ll 1$ is sufficiently small. The frequency content of the measured system response is accurate up to the Nyquist frequency $1/(2\Delta t)$.

In the following discussion, we usually require that $N_s \to \infty$ in the limit. We omit the limit notation for simplicity with an understanding that $N_s \gg 1$ is sufficiently large.

We assume that we don't have equations for the system and we only have data. The response moments can be computed from the measurements [1]. The mean, mean square and auto-correlation function of the response can be computed as

$$\mathrm{E}[x(t_k)] = \mu_x(t_k) = \frac{1}{N_s}\sum_{n=1}^{N_s} x_n(t_k), \tag{11.1}$$

$$\mathrm{E}[x^2(t_k)] = \frac{1}{N_s}\sum_{n=1}^{N_s} x_n^2(t_k), \tag{11.2}$$

$$\begin{aligned} \mathrm{E}[x(t_k)x(t_k+\tau)] &= R_{xx}(t_k, t_k+\tau) \\ &= \frac{1}{N_s}\sum_{n=1}^{N_s} x_n(t_k)x_n(t_k+\tau). \end{aligned} \tag{11.3}$$

Assume that $x(t_k)$ is a measurement of the response process that is ergodic in the mean and auto-correlation. N_s is the number of data points. Then, we have the mean, mean square and auto-correlation function of the response as

$$\mathrm{E}[x(t)] = \mu_x = \frac{1}{N_s}\sum_{k=1}^{N_s} x(t_k), \tag{11.4}$$

$$\mathrm{E}[x^2(t)] = \frac{1}{N_s}\sum_{k=1}^{N_s} x^2(t_k), \tag{11.5}$$

$$\begin{aligned} \mathrm{E}[x(t)x(t+\tau)] &= R_{xx}(\tau) \\ &= \frac{1}{N_s}\sum_{k=1}^{N_s} x(t_k)x(t_k+\tau). \end{aligned} \tag{11.6}$$

The statistical properties of the above averages can be studied as is commonly done in statistics.

If we take more than one measurements from a system, for example, $x_n(t_k)$ and $y_m(t_k)$, we can compute their means, second order moments and cross-correlations in the same manner. When the system has two or more outputs, it may have more than one degree of freedom. Hence, more data are needed. How many outputs we need to characterize a dynamic system without the knowledge of its mathematical model is related to the embedding dimension of the underlining system, which is discussed Chap. 12.

11.3 Density Estimation

Extensive random vibration measurements allow us to estimate the PDF of the measured output. At least, there are two ways to approach this problem. The first method makes use of response moments of various order. The second method involves two steps and can be discussed in the framework of generalized cell mapping. In this section, we discuss how to compute the PDF of the output by making use of response moments. The second method will be discussed in Chap. 12.

For brevity of discussion, we consider a single variable $x(t_k)$ which is a stationary process. Assume that we have computed its moments up to M^{th} order. The jth order moment is denoted as m^j. Assume that the PDF of x can be expressed as

$$p_x(x, \boldsymbol{\beta}) = C_0 e^{-\phi(x,\beta_j)}, \tag{11.7}$$

$$\phi(x, \boldsymbol{\beta}) = \sum_{j=1}^{M} \beta_j x^j, \tag{11.8}$$

where $\boldsymbol{\beta}$ are undetermined coefficients β_j of the PDF function $p_x(x, \boldsymbol{\beta})$, and C_0 is a normalization constant to be determined. We consider the following optimization problem to find $\boldsymbol{\beta}$ in order to minimize the error of moment prediction.

$$\min_{\boldsymbol{\beta}} \sum_{j=1}^{M} \left(\int_{R^1} x^j p_x(x, \boldsymbol{\beta}) dx - m^j \right)^2 \text{ subject to} \tag{11.9}$$

$$\int_{R^1} p_x(x, \boldsymbol{\beta}) dx = 1. \tag{11.10}$$

A popular choice of the unknown function $\phi(x, \beta_j)$ is a polynomial of x up to a certain order. The order can be determined based on the knowledge of the system. We can also select an order sufficiently high and apply a sparse algorithm such as SINDy to remove the unnecessary terms, as demonstrated in the following sections.

11.4 Model Identification with Data+Knowledge

Random vibration data are the best from the view point of system identification because such data contain information of the system dynamics in a wide range of frequencies. This is beneficial to determining the parameters of the system. In this section, we present an algebraic method to deal with random data when differentiations of the data are needed [46, 47].

As an example, we use a single-degree-of-freedom system to illustrate the research concept. Assume that the system with an inertial element of mass m satisfies Newton's second law such that the acceleration $\ddot{x}$ is governed by

$$m\ddot{x} + g(x, \dot{x}) = f(t), \tag{11.11}$$

where $f(t)$ the excitation and $g(x, \dot{x})$ represents the internal and nonlinear restoring force of the system. In general, we can express $g(x, \dot{x})$ in terms of a series expansion of regular or orthogonal polynomials. Because the range of the variables $(x, \dot{x})$ may not be bounded and is unknown, it is natural to consider the regular polynomials with time-invariant coefficients given by,

$$g(x, \dot{x}) = \sum_{i=0}^{N}\sum_{j=0}^{M} c_{ij} x^i \dot{x}^j, \tag{11.12}$$

where c_{ij} are the unknown coefficients of the polynomial. The coefficients c_{ij} are usually related to the physical properties of the dynamic system. For the system without any prior knowledge, the orders N and M of the polynomial are unknown. This study develops an algorithm to identify the polynomial defined in Eq. (11.12) from the input and output data such that it has a minimum number of terms and with the minimum orders N and M.

11.4.1 Algebraic Method

We prepare necessary signals from the measurements by using the algebraic method in order to apply the least square method to compute the parameters m and c_{ij} of the system [42]. We use a nonlinear oscillator to illustrate the steps of the method.

$$\begin{aligned} &m\ddot{x} + c_1\dot{x} + c_2\dot{x}x^2 + c_3\dot{x}^3 + k_1 x + k_3 x^3 = f(t), \\ &x(0) = x_0, \ \dot{x}(0) = v_0, \end{aligned} \tag{11.13}$$

where (x_0, v_0) are the initial conditions, which are generally unknown. We introduce several short-hands for the nonlinear terms.

$$z_1 = \dot{x}x^2, \quad z_2 = \dot{x}^3, \quad z_3 = x^3. \tag{11.14}$$

Applying the Laplace transform to Eq. (11.13), we have

$$\begin{aligned} &m\left\{s^2X(s) - sx_0 - v_0\right\} \\ &+ c_1\left\{sX(s) - x_0\right\} + k_1X(s) = G(s), \end{aligned} \tag{11.15}$$

where

$$G(s) = F(s) - c_2Z_1(s) - c_3Z_2(s) - k_3Z_3(s). \tag{11.16}$$

$X(s)$ is the Laplace transform of $x(t)$, $Z_i(s)$ is the Laplace transform of $z_i(t)$ and $F(s)$ is the Laplace transform of $f(t)$.

To eliminate the dependence on the unknown initial conditions x_0 and v_0, we take derivatives with respect to s twice in Eq. (11.15) and obtain

$$\begin{aligned} &m\left\{2X(s) + 4s\frac{dX(s)}{ds} + s^2\frac{d^2X(s)}{ds^2}\right\} \\ &+ c_1\left\{2\frac{dX(s)}{ds} + s\frac{d^2X(s)}{ds^2}\right\} + k_1\frac{d^2X(s)}{ds^2} = \frac{d^2G(s)}{ds^2}. \end{aligned} \tag{11.17}$$

Dividing both sides of Eq. (11.17) by s^2 to avoid derivatives with respect to time, we obtain

$$\begin{aligned} &m\left\{\frac{2}{s^2}X(s) + \frac{4}{s}\frac{dX(s)}{ds} + \frac{d^2X(s)}{ds^2}\right\} \\ &+ c_1\left\{\frac{2}{s^2}\frac{dX(s)}{ds} + \frac{1}{s}\frac{d^2X(s)}{ds^2}\right\} \\ &+ k_1\frac{1}{s^2}\frac{d^2X(s)}{ds^2} = \frac{1}{s^2}\frac{d^2G(s)}{ds^2}. \end{aligned} \tag{11.18}$$

Equation (11.18) can be rewritten in terms of the following signals

$$\begin{aligned} &mP_1(t) + c_1P_2(t) + k_1P_3(t) + \\ &c_2P_4(t) + c_3P_5(t) + k_3P_6(t) = P_f(t), \end{aligned} \tag{11.19}$$

where

$$
\begin{aligned}
P_1(t) &= 2\int^{(2)} x(t) - 4\int tx(t) + t^2x(t), \\
P_2(t) &= -2\int^{(2)} tx(t) + \int t^2x(t), \\
P_3(t) &= \int^{(2)} t^2x(t), \quad P_4(t) = \int^{(2)} t^2z_1(t), \\
P_5(t) &= \int^{(2)} t^2z_2(t), \quad P_6(t) = \int^{(2)} t^2z_3(t), \\
P_f(t) &= \int^{(2)} t^2 f(t).
\end{aligned}
\tag{11.20}
$$

Note that $\int^{(n)} \phi(t)$ denotes the multiple integral $\int_0^t \int_0^{\sigma_1} \cdots \int_0^{\sigma_{n-1}} \phi(\sigma_n)d\sigma_n \cdots d\sigma_1$. Let t_k $(k = 1, 2, \ldots, n_t)$ be a set of sampled times. Define an error at time t_k as

$$
\begin{aligned}
e(k) = {} & mP_1(t_k) + c_1P_2(t_k) + k_1P_3(t_k) \\
& + c_2P_4(t_k) + c_3P_5(t_k) + k_3P_6(t_k) - P_f(t_k).
\end{aligned}
\tag{11.21}
$$

Let us introduce an $n_t \times 1$ error vector as $\mathbf{e} = [e(1), e(2), \ldots, e(n_t)]^T$, a parameter vector as $\mathbf{c} = [m, c_1, k_1, c_2, c_3, k_3]^T$ and a force vector as $\mathbf{p} = [P_f(t_1), P_f(t_2), \ldots, P_f(t_{n_t})]^T$. Let $\mathbf{P}(x)$ be the $n_t \times 6$ data matrix determined by the measurement of $x(t)$ such that its $(k, j)^{th}$ component is given by $P_{k,j} = P_j(t_k)$. Then, Eq. (11.21) can be written in the matrix form as

$$
\mathbf{e}(\mathbf{c}) = \mathbf{P}(x)\mathbf{c} - \mathbf{p}. \tag{11.22}
$$

To find the unknown parameters $\mathbf{c}$, we formulate a least mean square error problem as follows. For the following cost function,

$$
J(\mathbf{c}) = \mathbf{e}^T\mathbf{e} = \|\mathbf{e}\|_2^2, \tag{11.23}
$$

we find the parameter vector $\mathbf{c}$ such that J is minimized. The solution of the parameters $\mathbf{c}$ can be found by either the matrix inversion or by an iterative search algorithm. In general, when the numbers of unknown parameters and data points n_t are large, as is the case for multi-degree of freedom systems, it is better to use a global search algorithm to compute the parameters.

11.4.2 Computation of Multiple Integration

Let us consider one example of the integrals that define the signals.

$$\int^{(2)} tx(t) = \int_0^t \int_0^T t_1 x(t_1) dt_1 d\mathcal{T} \equiv y_1(t). \tag{11.24}$$

Then, we have

$$\dot{y_1}(t) = \int_0^t t_1 x(t_1) dt_1 \equiv y_2(t),$$
$$\dot{y_2}(t) = tx(t). \tag{11.25}$$

Hence, we arrive at two linear state equations with $tx(t)$ as input to the system

$$\dot{y_1}(t) = y_2(t),$$
$$\dot{y_2}(t) = tx(t). \tag{11.26}$$

The output of the state equations is $y_1(t)$, which is the double integration of the input. It is interesting to point out that the initial conditions for the states are always zero.

$$y_1(0) = 0, \quad y_2(0) = 0. \tag{11.27}$$

Hence, to compute the multiple integration of a signal, we can numerically integrate the linear state equations starting from zero initial conditions with the signal as input.

11.4.3 Sparse Representation

Assume that the system modeled by Eq. (11.13) is unknown and that only the displacement measurement $x(t)$ of the system is available and contains random noises. We consider the polynomial model for the restoring force in Eq. (11.12). The questions are:

1. What are the minimum orders N and M?
2. Are all the terms in the polynomial needed?

One popular way is to start with high enough orders of the polynomial or a large library of functions, which may contain non-polynomial functions, and penalize the the terms with smaller coefficients through a sparse regulator by using the least absolute shrinkage and selection operator (LASSO) [57]. The recent studies of the sparse identification of nonlinear dynamics (SINDy) have developed a computationally efficient algorithm to solve the sparse regression. To find the sparse regulator, SINDy uses the "elbow" of the bias-variance tradeoff curve called the "Pareto front". In most of cases, finding the curve's elbow is ambiguous due to existence of a cluster of candidate models close to the elbow. To address this issue, an information criterion for the candidate models can be used to rank and select the right model [49]. It is important to point out that the selection of the models on the Pareto front from a large library can be computationally intensive.

Here, we build the models starting from lower order polynomials. We divide the data into the training set and test set. In the training, we apply the sparse regression LASSO to select the models and employ the K-fold cross validation within the training set to find the sparse regulator. This process generates a model with the smallest mean squared cross validation error in the training set for a given order of the polynomial. Finally, we compute the information criterion of the trained model over the test set, and then increase the order of the polynomial. We stop increasing the order of the polynomial after the information criterion reaches a minimum value and starts to grow.

For a given order of the polynomial, the LASSO proposes to add an L^1 regularization term to the cost function in Eq. (11.23) in order to penalize the nonzero parameters that don't contribute much to the model [57].

$$
\begin{aligned}
J_\lambda(\mathbf{c}) &= \|\mathbf{e}\|_2^2 + \lambda \|\mathbf{c}\|_1 \\
&= \|\mathbf{P}(x)\mathbf{c} - \mathbf{p}\|_2^2 + \lambda \|\mathbf{c}\|_1 ,
\end{aligned}
\tag{11.28}
$$

where $\lambda > 0$ is a preselected small positive number known as the sparse regulator and $\|\mathbf{c}\|_1 = \sum_i |c_i|$. The sparse regression algorithm in [48] is applied to find the parameter vector $\mathbf{c}$ in order to minimize J_λ in Eq. (11.28). The algorithm is computationally efficient and robust to noise. It should be noted that the optimization problem in the framework of the LASSO is convex [17], which implies the existence of a unique optimal solution.

In the studies of SINDy, the sparse regulator with parameter λ is determined through cross validation [48]. The simulations from different initial conditions are usually used to generate rich data sets to train and validate the model. The training data is often much larger than the validation data. Although the rich data set leads to more accurate model with the help of the regularization term, especially for stochastic systems, it is quite expensive to generate such a data set from experiments. To resolve this issue, we apply the K-fold cross validation method by splitting the training set of a long time series of the measurement $x(t)$ into K roughly equal parts or folds as Fig. 11.1 illustrates. One of the K folds of the training dataset is used for cross validation of the model trained with the rest of the data in the same training dataset.

Let T_k denote the time instances of the kth part of the K folds to be used for validation, and $T_{\bar{k}}$ denote its compliment, i.e. the time instances of the rest of the K folds of the data to be used for training. The model is trained on each fold of the data in $X_{\bar{k}}$ and is validated on the set X_k. For a given regulation parameter λ, the

Training	Training	.....	Validation	.....	Training	Training
1	2	.....	*k*	.....	*K-1*	*K*

Fig. 11.1 Illustration of the data segmentation for the K-fold cross validation

mean square cross validation error MSE_{CV} of the model on all the validation sets is defined as

$$MSE_{CV}(\lambda) = \frac{1}{n_t} \sum_{k=1}^{K} \|\mathbf{P}(x(t \in T_k))\mathbf{c}_k - \mathbf{p}(t \in T_k)\|_2^2, \tag{11.29}$$

where $\mathbf{c}_k$ denote all the model parameters trained on $T_{\bar{k}}$. MSE_{cv} is an implicit function of λ. The minimum MSE_{cv} with respect to λ represents a compromise between the accuracy and complexity of the model [48]. This leads to the optimal choice of the regulation parameter λ for the given order of the polynomial.

Next, we consider the variation of the sparse representation of the candidate model with the order of the polynomial. Here, we adopt an information criterion to rank the sparse representation of the model [49]. Popular statistical examples of information criteria include the Akaike information criterion (AIC), Bayesian information criterion (BIC), deviance information criterion (DIC) and minimum description length. This work considers the AIC for evaluating the candidate models. The AIC describes the generality and goodness of fit of the candidate models and is calculated on the test dataset that is not used in training.

Let n_c denote the number of nonzero parameters in the vector $\mathbf{c}$. The AIC index for the model is defined as

$$AIC = 2n_c - 2\ln(L(x, \mathbf{c})), \tag{11.30}$$

where $L(x, \mathbf{c}) = p(x|\mathbf{c})$ is the likelihood function of the prediction x by the model with the parameters $\mathbf{c}$. Assume that the error $\mathbf{e}$ which contains the measurement noise is normally distributed, the residual sum of squares (RSS) can be considered as the likelihood function such that the AIC for the model can be defined as [58]

$$AIC = n_t \ln(RSS/n_t) + 2n_c. \tag{11.31}$$

If the least squares estimation is used, the RSS is equal to the cost function $RSS = J(\mathbf{c})$ in Eq. (11.23). We summarize the previous steps in Algorithm 2.

Algorithm 2 The proposed algorithm for nonlinear system identification with the algebraic method.

procedure SYSTEM IDENTIFICATION $(x, f, k_{fold}, ratio, \lambda_{min}, \lambda_{max})$
  for $PolyOrder \in [1, N;\ 1, M]$ **do**
    $g(x, \dot{x}) \leftarrow x, \dot{x}, PolyOrder$ ▷ Generate polynomial terms
    $P(X) \leftarrow \text{library}(X, PolyOrder)$
    $\mathbf{P}(x) \leftarrow \text{library}(\text{Algebraic Operation}(g(x, \dot{x})))$ ▷ Add new polynomial terms only
    $\mathbf{p} \leftarrow \text{Algebraic Operation}(f(t))$
    $\mathbf{P}_{train}(x), \mathbf{p}_{train}, \mathbf{P}_{test}(x), \mathbf{p}_{test} \leftarrow ratio$ ▷ Split data for training and test
    for $\lambda \in [\lambda_{min}, \lambda_{max}]$ **do** ▷ Search for the sparse regulator
      $MSE_{cv}(\lambda) \leftarrow \text{SINDy \& K} - \text{fold}(\mathbf{P}_{train}(x), \mathbf{p}_{train}, \lambda)$
    $\lambda_{cv,min} \leftarrow min(MSE_{cv}(\lambda))$
    $\text{Model}(PolyOrder) \leftarrow \text{SINDy}(\mathbf{P}_{train}(x), \mathbf{p}_{train}, \lambda_{cv,min})$ ▷ Sparse models
  $AIC(PolyOrder) \leftarrow \text{AIC}(\text{Model}, \mathbf{P}_{test}(x), \mathbf{p}_{test})$ ▷ AIC of candidate models
  $\text{Model}(\mathbf{c}) \leftarrow \text{Overall } min(AIC)$ ▷ Select the model with minimum AIC
  return Model ($\mathbf{c}$)

11.4.4 Numerical Example

To validate the proposed algorithm, we consider a second-order oscillatory system with nonlinear stiffness and damping.

$$m\ddot{x} + k_1 x + k_3 x^3 + c_1 \dot{x} + c_2 \dot{x} x^2 + c_3 \dot{x}^3 = f(t), \tag{11.32}$$

where the external force $f(t)$, as an example, is given by

$$f(t) = 10\sin(t) + 40\sin(4t) + \epsilon_f. \tag{11.33}$$

ϵ_f is a normally distributed Gaussian noise with zero mean and standard deviation $\sigma_{\epsilon_f} = 0.01$. In the simulations, we have set $k_1 = 6$, $k_3 = 0.1$, $c_1 = 1$, $c_2 = 0.2$ and $c_3 = 1$. A measurement noise is considered such that the system output is given by $x(t) + \epsilon_x$ where ϵ_x is another normally distributed random noise with zero mean and standard deviation $\sigma_{\epsilon_x} = 0.01$. The general restoring force function $g(x, \dot{x})$ in Eq. (11.12) is used to model the system.

We follow the steps in Algorithm 2 to identify the best model for the data. The order of the polynomial starts from 1 and increases to 6. For simplicity, we have taken $N = M$ in the simulations. The time series generated from Eq. (11.32) is equally divided into the training and test datasets. We choose $K = 5$ for the K-fold cross validation to train and select the candidate model by using the training dataset. The sparse representation of the model with a given order of the polynomial is further tested by using the test dataset. On the test dataset, the AIC index is also computed.

The results of the study are well summarized in Figs. 11.2 and 11.3. Figure 11.2 shows the variation of the AIC index of the sparse model as a function of the order of

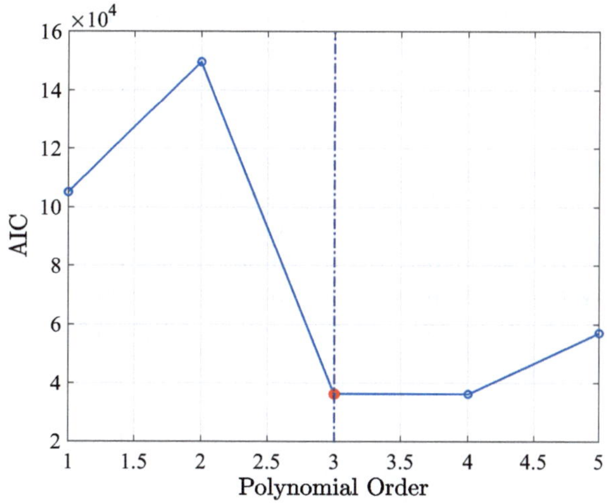

Fig. 11.2 The AIC index of the model as a function of the order of polynomial for the system in Eq. (11.32) when $\lambda_{\min}$ is used for all the models

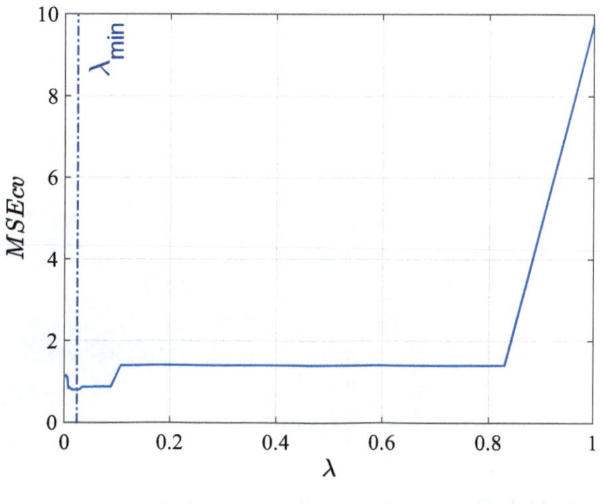

Fig. 11.3 The mean squared error MSE_{cv} of the K-fold cross validation as a function of the regulation parameter λ for the system in Eq. (11.32) when the model is the third order polynomial

the polynomial. The curve indicates that the AIC index reaches its minimum when the order is 3. This is consistent to the known order of the system.

Figure 11.3 shows the variation of the mean squared error MSE_{cv} of the K-fold cross validation for the model with the third order polynomial with the regulation parameter λ. The figure clears shows that there exists an optimal λ such that MSE_{cv} is minimum. This is true for all the orders of the polynomials we computed.

The resulting model is optimal in the sense of sparse representation of the SINDy. Table 11.1 compares the true parameters of the system with the estimated values of the parameters when the polynomial of the model is third order $N = M = 3$ when the AIC index is minimum. It is evident from the results that the proposed method for identifying the model of nonlinear dynamic systems is quite accurate and effective.

Table 11.1 The true and estimated coefficients of the polynomial terms in Eq. (11.32)

Terms	True value	Estimated	Terms	True value	Estimated
$\ddot{x}$	2	1.9981	x	6	5.9857
x^2	0	0	x^3	0.1	0.1059
$\dot{x}$	1	0.9147	$\dot{x}^2$	0	0
$\dot{x}^3$	1	1.0064	$x\dot{x}$	0	0
$x\dot{x}^2$	0	0	$x\dot{x}^3$	0	0
$x^2\dot{x}$	0.2	0.1973	$x^2\dot{x}^2$	0	0
$x^2\dot{x}^3$	0	0	$x^3\dot{x}$	0	0
$x^3\dot{x}^2$	0	0	$x^3\dot{x}^3$	0	0

Fig. 11.4 Quanser rotary flexible beam set up

11.4.5 Experimental Example

Here, we apply the proposed algorithm to identify the model of a multiple degree of freedom (MODF) system from experimental data. In particular, we consider a rotary flexible beam made by Quanser in the experimental study. Figure 11.4 depicts the setup. The flexible beam is driven by a DC servomotor with an input voltage V_m. The angle of the base holding the flexible beam is θ. The deflection angle of the beam is denoted as α. Table 11.2 lists the system parameters.

The beam has a moment of inertia J_l about the center of rotation and a mass m_l. It is rigidly fixed at one end and mounted on the servo motor. The servo motor and gear

Table 11.2 The parameters of the rotary flexible beam provided by Quanser [59]

Symbol	Description	Value
J_l	Rotational moment of inertia	$0.0038\,\mathrm{kg}\cdot\mathrm{m}^2$
J_{eq}	Bending moment of inertia	$9.76 * 10^{-5} kg.m^2$
B_l	Damping coefficient	Unknown
B_{eq}	Viscous damping	$0.015\,\mathrm{kg}$
k_s	Beam stiffness	$1.3\ \frac{N\cdot m}{rad}$
k_g	Gear ratio	70
k_t	Motor current torque constant	$7.68 * 10^{-3} \frac{N\cdot m}{A}$
k_m	Motor back-emf constant	$7.68 * 10^{-3} \frac{v}{rad/s}$
R_m	Motor armature resistance	2.6Ω
η_g	Gearbox efficiency	0.9
η_m	Motor efficiency	0.69

system have Coulomb friction with an unknown coefficient B_c and viscous damping coefficient B_{eq}. The beam has viscous damping due to the air with the coefficient B_l. The flexible beam has a linear stiffness K_s when the motion is small, and the stiffness becomes nonlinear when the excitation level is high.

The equations of motion for the system can be derived as [60],

$$\begin{aligned}
&(J_{eq} + J_l)\ddot{\theta} + J_l\ddot{\alpha} + B_{eq}\dot{\theta} = \tau, \\
&J_l\ddot{\alpha} + J_l\ddot{\theta} + B_l\dot{\alpha} + k_s\alpha = 0, \\
&\tau = \frac{\eta_g k_g \eta_m k_t (V_m - k_g k_m \dot{\theta})}{R_m}.
\end{aligned} \tag{11.34}$$

The system has two outputs θ and α under the input of voltage V_m. We assume that the torque τ is linearly dependent on voltage V_m, while the parameters of the motor are unknown. Table 11.2 shows the parameters provided by Quanser.

In terms of the control voltage, the equations of motion (11.34) can be written as

$$\begin{aligned}
&\frac{R_m}{\nu_g k_g \eta_m k_t}\left[J_{eq}\ddot{\theta} + \left(B_{eq} + \frac{\nu_g k_g^2 k_m \eta_m k_t}{R_m}\right)\dot{\theta} - k_s\alpha\right] = V_m, \\
&-\frac{R_m}{\nu_g k_g \eta_m k_t} \\
&\times\left[J_{eq}\ddot{\alpha} - \left(\frac{\nu_g k_g^2 k_m \eta_m k_t}{R_m} + B_{eq}\right)\dot{\theta} + k_s\frac{J_l + J_{eq}}{J_l}\alpha\right] = V_m.
\end{aligned} \tag{11.35}$$

We should point out that the original model of the system (11.35) does not account for the effect of friction and nonlinearity, which exist in the experimental data. We assume that the restoring force of the system (11.35) can be represented by a sum of four polynomials of a single variable, i.e. θ^i, $\dot{\theta}^i$, α^j, and $\dot{\alpha}^j$, together with a friction term due to $\dot{\theta}$, without considering the cross-product terms.

Some signals contain the terms involving the first order derivative of the angle such as $\dot{\theta}^i$. We use the nonlinear state observer from the control studies to estimate the first-order derivatives. Let us take θ as an example. The second order observer with the gains β_1 and β_2 can be implemented with the following state equations.

$$\dot{z}_1 = z_2 - \beta_1 e,$$
$$\dot{z}_2 = -\beta_2 \text{fal}(e, \alpha, \delta), \tag{11.36}$$

where

$$\text{fal}(e, \alpha, \delta) = \begin{cases} |e|^\alpha \cdot \text{sign}(e), & |e| > \delta, \\ \frac{e}{\delta^\alpha}, & |e| \le \delta, \end{cases} \tag{11.37}$$

and the error $e = z_1 - \theta$ stands for the difference between θ and its observer estimate z_1. We have chosen $\alpha = 0.5$, $\delta = 0.05$, $\beta_1 = 100$ and $\beta_2 = 900$. According to Guo and Zhao [61], Han [62], the estimation error e converges to zero quickly. Consequently, by definition, z_2 is an accurate estimate of the first derivative $\dot{\theta}$. We split the data into two equal sets for training and test. We chosen $K = 10$ for the K-fold cross validation during the training. We have considered the order of the polynomials from 1 up to 4.

We again follow the steps in Algorithm 2 to identify the best model for the data. The results are summarized in Figs. 11.5 and 11.6. Figure 11.5 shows the AIC indices for both the angles as a function of the order of the polynomials when the optimal regulation parameters λ_θ and λ_α are used. The figure clearly indicates that the optimal order of the polynomial for both the angles is three. This is consistent with the geometrical nonlinearity of the beam, which is typically the third order in the displacement variable.

Figure 11.6 shows the mean squared error MSE_{cv} of the K-fold cross validation for the third order polynomial model as a function of the regulation parameters λ_θ and λ_α for the system in Eq. (11.35). The figure indicates the optimal values of the regulation parameters λ_θ and λ_α for both the angles α and θ.

Figures 11.5 and 11.6 indicate the existence of an optimal model from experimental data. The following equations of motion are the optimal polynomial model of the system with the smallest AIC indices: $AIC_\theta = -1.0205 \times 10^4$ and $AIC_\alpha = -8.833 \times 10^3$.

$$a_{11}\ddot{\theta} + a_{12}\dot{\theta} + a_{13}\theta + a_{14}\theta^2 + a_{15}\theta^3$$
$$+ a_{16}\dot{\alpha} + a_{17}\alpha + a_{18}\alpha^2 + a_{19}\text{sign}(\dot{\theta}) = V_m, \tag{11.38}$$
$$a_{21}\ddot{\theta} + a_{22}\dot{\theta} + a_{23}\theta + a_{24}\theta^2 + a_{25}\theta^3$$
$$+ a_{26}\dot{\alpha} + a_{27}\alpha + a_{28}\alpha^2 + a_{29}\text{sign}(\dot{\theta}) = V_m, \tag{11.39}$$

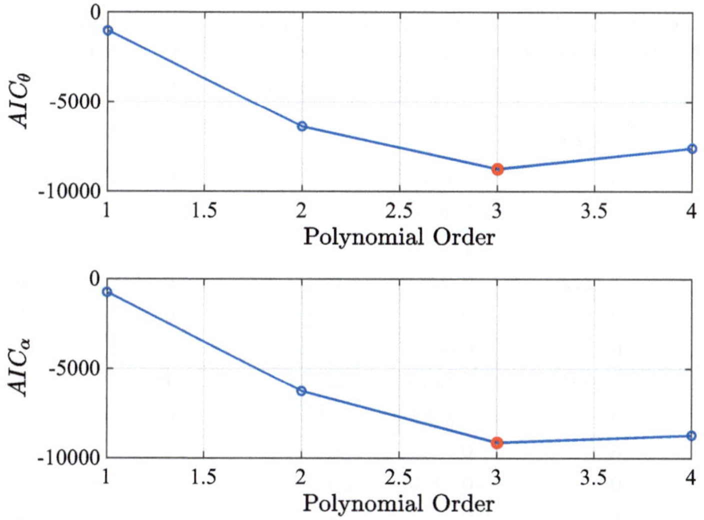

Fig. 11.5 The AIC index of the model as a function of the order of polynomial for the rotatory flexible beam system in Eq. (11.35) when $\lambda_{\theta,\min}$ and $\lambda_{\alpha,\min}$ are used for all the models

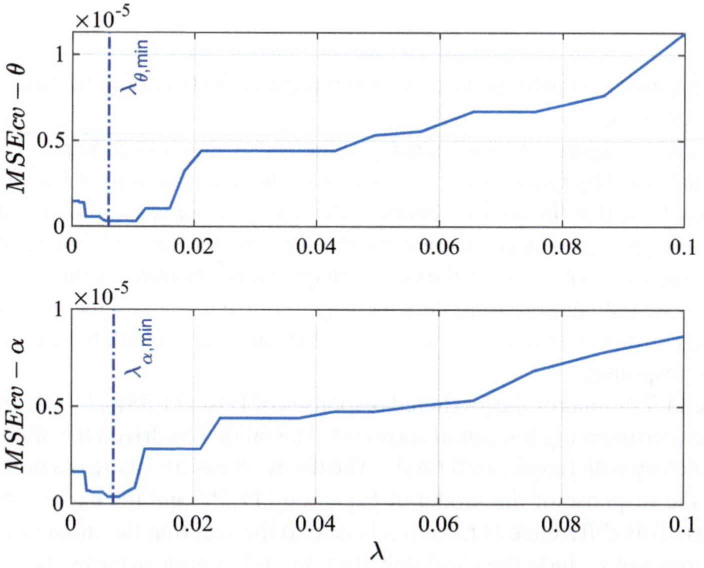

Fig. 11.6 The mean squared error MSE_{cv} of the K-fold cross validation as a function of the regulation parameters λ_θ and λ_α for the rotatory flexible beam system in Eq. (11.35) when the model consists of the third order polynomials

Table 11.3 The estimated coefficients of the polynomial terms in Eqs. (11.38) and (11.39) for the rotary flexible listed to compare with the identified numbers. f_c is the coefficient of friction

Terms in Eq. (11.38)	Quanser	Identified	Terms in Eq. (11.39)	Quanser	Identified
$\ddot{\theta}$	0.016231	0.0190	$\ddot{\alpha}$	−0.016231	−0.0118
θ	0	0.4151	θ	0	0.7113
θ^2	0	0.0933	θ^2	0	0.0935
θ^3	0	−1.5354	θ^3	0	−2.7442
$\dot{\theta}$	0.7323	0.6623	$\dot{\theta}$	0.7323	0.5890
$\dot{\theta}^2$	0	0	$\dot{\theta}^2$	0	0
$\dot{\theta}^3$	0	−0.0065	$\dot{\theta}^3$	0	0
α	−10.124	−6.3840	α	−15.672	−13.1712
α^2	0	5.8953	α^2	0	6.2922
α^3	0	0	α^3	0	0
$\dot{\alpha}$	0	−0.0420	$\dot{\alpha}$	0	−0.0872
$\dot{\alpha}^2$	0	0	$\dot{\alpha}^2$	0	0
$\dot{\alpha}^3$	0	0	$\dot{\alpha}^3$	0	0
f_c	0	0.1151	f_c	0	0.1474

where the terms $a_{19}\text{sign}(\dot{\theta})$ and $a_{29}\text{sign}(\dot{\theta})$ represent the Coulomb friction that exists in the experiment.

Table 11.3 compares the estimated parameters of Eqs. (11.38) and (11.39) with the ones provided by Quanser corresponding to the linear system of Eq. (11.35). We should point out that the parameters provided by Quanser are nominal and may not represent the physical system accurately. If the nonlinear Eqs. (11.38) and (11.39) of the identified model represents the physical system better in real time, the estimated parameters would be more reliable. One way to make this assessment is to compare the predicted responses of Eqs. (11.35), (11.38) and (11.39) with the experimentally measured response.

Figure 11.7 compares the predicted responses of Eqs. (11.35), (11.38) and (11.39) with the experimentally measured response. The angle θ is driven to follow a square wave trajectory with frequency 0.66 Hz. The figure shows that there is a big difference between the response of the model of Equation (11.35) and the experimental measurements. This difference is most likely due to the fact that the model of Equation (11.35) does not include the Coulomb friction and nonlinear terms, both of which exit in the experimental data.

On the other hand, Eqs. (11.38) and (11.39) contain both the Coulomb friction and nonlinear terms. The Coulomb friction coefficients are estimated to be $a_{19} = 0.1151$ and $a_{29} = 0.1474$. The coefficients of nonlinear terms, e.g. a_{12} and a_{22}, are considerable. Generally, the estimated value of the coefficients of the linear terms are close to the nominal values provided by Quanser.

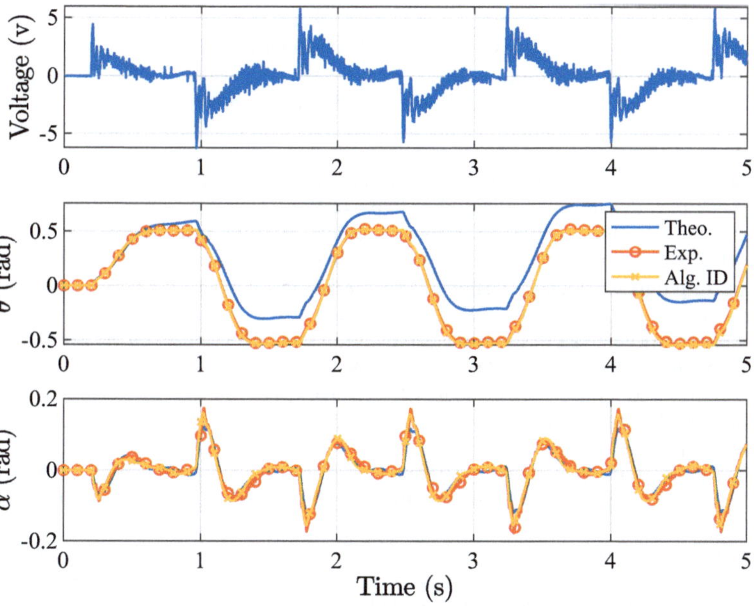

Fig. 11.7 Comparison of the responses of Eqs. (11.35), (11.38) and (11.39) with the experimentally measured response. The response of θ follows a square trajectory with the amplitude of 30° and frequency 0.66 Hz. Top: Noisy Input voltage. Middle: θ angle response. Bottom: α angle response. Solid line: Eq. (11.35). Solid line+circle: Experimental data. Solid line+cross: Eqs. (11.38) and (11.39)

References

1. Bendat JS, Piersol AG (2000) Random data: analysis and measurement procedures. Wiley, New York
2. Li Z, Jiang J, Hong L, Sun JQ (2019) On the data-driven generalized cell mapping method. Int J Bifurc Chaos 29(14):1950,204. https://doi.org/10.1142/S0218127419502043
3. Li Z, Qiao L, Jiang J, Hong L, Sun JQ (2020) Global dynamic analysis of the north pacific ocean by data-driven generalized cell mapping method. Int J Dyn Control 8(4):1141–1146. https://doi.org/10.1007/s40435-020-00678-z
4. Ma SF, Leylaz G, Sun JQ (2022) Identification of differentially flat output of underactuated dynamic systems. Int J Control 95(1):114–125. https://doi.org/10.1080/00207179.2020.1779960
5. Ma SF, Leylaz G, Sun JQ (2022) Data-driven robust tracking control of underactuated mechanical systems using identified flat output and active disturbance rejection control. Int J Control 95(11):3101–3117
6. Gautrais J, Ginelli F, Fournier R, Blanco S, Soria M, Hugues C, Guy T (2012) Deciphering interactions in moving animal groups. PLOS Comput Biol
7. Zienkiewicz A, Barton DAW, Porfiri M, di Bernardo M (2015) Data-driven stochastic modelling of zebrafish locomotion. J Math Biol 71(5):1081–1105
8. Burbano L, Daniel A, Porfiri M (2020) Data-driven modeling of zebrafish behavioral response to acute caffeine administration. J Theor Biol 485(110):054

9. Hernandez AM, Casado Magana EJ, Berna AG (2018) Data-driven aircraft trajectory predictions using ensemble meta-estimators. In: Proceedings of the 37th digital avionics systems conference, London, UK, pp 1–10
10. Altchë F, de La Fortelle A (2017) An LSTM network for highway trajectory prediction. In: Proceedings of the 20th International Conference on Intelligent Transportation Systems, Yokohama, Japan, pp 353–359
11. Wang WX, Lai YC, Grebogi C (2016) Data based identification and prediction of nonlinear and complex dynamical systems. Phys Rep 644:1–76
12. Ljung L (2010) Perspectives on system identification. Annu Rev Control 34(1):1–12
13. Noël JP, Kerschen G (2017) Nonlinear system identification in structural dynamics: 10 more years of progress. Mech Syst Signal Process 83:2–35
14. Schoukens J, Vaes M, Pintelon R (2016) Linear system identification in a nonlinear setting: nonparametric analysis of the nonlinear distortions and their impact on the best linear approximation. IEEE Control Syst Mag 36(3):38–69
15. Sracic MW, Allen MS (2014) Identifying parameters of multi-degree-of-freedom nonlinear structural dynamic systems using linear time periodic approximations. Mech Syst Signal Process 46(2):325–343
16. Moaveni B, Asgarieh E (2012) Deterministic-stochastic subspace identification method for identification of nonlinear structures as time-varying linear systems. Mech Syst Signal Process 31:40–55
17. Lai Z, Nagarajaiah S (2019) Sparse structural system identification method for nonlinear dynamic systems with hysteresis/inelastic behavior. Mech Syst Signal Process 117:813–842
18. Haroon M, Adams DE (2009) A modified H2 algorithm for improved frequency response function and nonlinear parameter estimation. J Sound Vib 320(4):822–837
19. Magnevall M, Josefsson A, Ahlin K, Broman G (2012) Nonlinear structural identification by the "Reverse Path" spectral method. J Sound Vib 331(4):938–946
20. Vazquez Feijoo JA, Worden K, Montes Garcia P, Lagunez Rivera L, Juarez Rodriguez N, Pech Pérez A (2010) Analysis of MDOF nonlinear systems using associated linear equations. Mech Syst Signal Process 24(8):2824–2843
21. Noël JP, Kerschen G (2013) Frequency-domain subspace identification for nonlinear mechanical systems. Mech Syst Signal Process 40(2):701–717
22. Noël JP, Renson L, Grappasonni C, Kerschen G (2016) Identification of nonlinear normal modes of engineering structures under broadband forcing. Mech Syst Signal Process 74:95–110
23. Clement S, Bellizzi S, Cochelin B, Ricciardi G (2014) Sliding window proper orthogonal decomposition: application to linear and nonlinear modal identification. J Sound Vib 333(21):5312–5323
24. Worden K, Green PL (2017) A machine learning approach to nonlinear modal analysis. Mech Syst Signal Process 84:34–53
25. Noël JP, Marchesiello S, Kerschen G (2014) Subspace-based identification of a nonlinear spacecraft in the time and frequency domains. Mech Syst Signal Process 43(1):217–236
26. Peng ZK, Lang ZQ, Wolters C, Billings SA, Worden K (2011) Feasibility study of structural damage detection using NARMAX modelling and nonlinear output frequency response function based analysis. Mech Syst Signal Process 25(3):1045–1061
27. Pai PF, Palazotto AN (2008) Detection and identification of nonlinearities by amplitude and frequency modulation analysis. Mech Syst Signal Process 22(5):1107–1132
28. Brewick PT, Masri SF, Carboni B, Lacarbonara W (2016) Data-based nonlinear identification and constitutive modeling of hysteresis in NiTiNOL and Steel strands. J Eng Mech 142(12):04016,107
29. Lusch B, Kutz JN, Brunton SL (2018) Deep learning for universal linear embeddings of nonlinear dynamics. Nat Commun 9(1):4950
30. Champion K, Lusch B, Kutz JN, Brunton SL (2019) Data-driven discovery of coordinates and governing equations. Proc Natl Acad Sci 116(45):22,445

31. Raissi M, Perdikaris P, Karniadakis G (2018) Multistep neural networks for data-driven discovery of nonlinear dynamical systems. arXiv:1801.01236
32. Billings SA (2013) Nonlinear system identification: NARMAX methods in the time, frequency, and spatio-temporal domains. Wiley, United States
33. Fu L, Li P (2013) The research survey of system identification method. In: Proceedings of the 5th international conference on intelligent human-machine systems and cybernetics, Hangzhou, China, vol 2, pp 397–401
34. Vlachas PR, Byeon W, Wan ZY, Sapsis TP, Koumoutsakos P (2018) Data-driven forecasting of high-dimensional chaotic systems with long short-term memory networks. Proc R Soc A: Math, Phys Eng Sci 474(2213)
35. Pathak J, Hunt B, Girvan M, Lu Z, Ott E (2018) Model-free prediction of large spatiotemporally chaotic systems from data: a reservoir computing approach. Phys Rev Lett 120(2):024,102
36. Muto M, Beck JL (2008) Bayesian updating and model class selection for Hysteretic structural models using stochastic simulation. J Vib Control 14(1–2):7–34
37. Beck JL, Katafygiotis LS (1998) Updating models and their uncertainties. I: Bayesian statistical framework. J Eng Mech 124(4):455–461
38. Green PL, Worden K (2015) Bayesian and Markov chain Monte Carlo methods for identifying nonlinear systems in the presence of uncertainty. Proc R Soc Math Phys Eng Sci 373(20140):405
39. Kerschen G, Golinval JC, Hemez FM (2003) Bayesian model screening for the identification of nonlinear mechanical structures. J Vib Acoust 125(3):389–397
40. Yosida K (1984) Operational calculus - a theory of hyperfunctions. Springer Science+Business Media, New York
41. Sira-Ramìrez H, Garcìa-Rodrìguez C, Cortés-Romero J, Luviano-Juárez A (2014) Algebraic identification and estimation methods in feedback control system. Wiley, United States
42. Fliess M, Sira-Ramìrez H (2003) An algebraic framework for linear identification. ESAIM: Control Optim Calc Var 9:151–168
43. Garrido R, Concha A (2013) An algebraic recursive method for parameter identification of a servo model. IEEE/ASME Trans Mechatron 18(5):1572–1580
44. Trapero JR, Sira-Ramìrez H, Batlle VF (2007) A fast on-line frequency estimator of lightly damped vibrations in flexible structures. J Sound Vib 307(1):365–378
45. Becedas J, Mamani G, Feliu-Batlle V, Sira-Ramìrez H (2007) Algebraic identification method for mass-spring-damper system. In: Proceedings of world congress on engineering and computer science, San Francisco, USA
46. Leylaz G, Ma SF, Sun JQ (2021) An optimal model identification algorithm of nonlinear dynamical systems with the algebraic method. J Vib Acoust 143:021,002–1–021,002–8
47. Leylaz G, Wang S, Sun JQ (2022) Identification of nonlinear dynamical systems with time delay. Int J Dyn Control 10:13–24. https://doi.org/10.1007/s40435-021-00783-7
48. Brunton SL, Proctor JL, Kutz JN (2016) Discovering governing equations from data by sparse identification of nonlinear dynamical systems. Proc Natl Acad Sci 113(15):3932
49. Mangan N, Kutz J, Brunton S, Proctor J (2017) Model selection for dynamical systems via sparse regression and information criteria. Proc R Soc Math Phys Eng Sci 473(20170):009
50. Kaiser E, Kutz J, Brunton S (2017) Sparse identification of nonlinear dynamics for model predictive control in the low-data limit. Proc R Soc Math Phys Eng Sci 474(20180):335
51. Mangan N, Askham T, Brunton S, Kutz J, Proctor J (2019) Model selection for hybrid dynamical systems via sparse regression. Proc R Soc Math Phys Eng Sci 475(20180):534
52. Chartrand R (2011) Numerical differentiation of noisy, nonsmooth data. ISRN Appl Math J 2011(161):561
53. Rudin LI, Osher S, Fatemi E (1992) Nonlinear total variation based noise removal algorithms. Phys D 60:259–268
54. Schaeffer H, McCalla SG (2017) Sparse model selection via integral terms. Phys Rev E 96(2):023,302

55. Zhao A, Huang J, Sun JQ (2023) Estimation of wheel-rail structural interactions from motion signals of high-speed train bogie. Int J Dyn Control 11:1609–1620. https://doi.org/10.1007/s40435-022-01085-2
56. Zhao A, Sun JQ (2023) Wear estimation of high speed train from motion measurements. J Vib Test Syst Dyn 7(3):307–326. https://doi.org/10.5890/JVTSD.2023.09.005
57. Hastie T, Tibshirani R, Friedman J (2009) The elements of statistical learning: data mining, inference, and prediction. Springer series in statistics
58. Burnham KP, Anderson DR (2002) Model selection and multi-model inference. Springer, Berlin, Germany
59. Apkarian J, Lévis M, Gurocak H (2011) User manual SRV02 rotary servo base unit set up and configuration. Quanser Inc., Markham, Ontario
60. Apkarian J, Karam P, Lévis M (2011) Flexible link experiment for MatLab/Simulink users. Quanser Inc., Markham, Ontario, Canada
61. Han J (1995) A class of extended state observers for uncertain systems. Control Decis 10:85–88
62. Guo B, Zhao ZL (2016) Active disturbance rejection control for nonlinear systems: an introduction. Wiley

Chapter 12
Data-Driven Generalized Cell Mapping

12.1 Generalized Cell Mapping

As pointed in Sect. 7.2.4, the generalized cell mapping (GCM) method has been applied to study random vibration problems of linear and nonlinear dynamic systems extensively. The cell mapping method proposes to discretize the state space into a collect of small hyper-boxes, called cells. Each cell is labeled by an integer. Evolution of the probability of the system response being found in a cell i at the beginning of a mapping step to its image cell j is determined by a finite Markov chain.

Let $\mathbf{p}(k) = \{p_j(k)\}$ be a vector of probabilities $p_j(k)$ that the system resides in the jth cell at the kth step. $\mathbf{P}(k) = \{p_{ij}(k)\}$ is the one step transition probability from the jth cell to i^{th} cell at the kth step. N_t is the total number of cells in the computational domain. We have

$$p_i(k+1) = \sum_{j=1}^{N_t} p_{ij}(k)p_j(k), \text{ for } k \geq 0. \tag{12.1}$$

In the vector form, it reads

$$\mathbf{p}(k+1) = \mathbf{P}(k)\mathbf{p}(k), \text{ for } k \geq 0. \tag{12.2}$$

When the response is a stationary stochastic process, the matrix $\mathbf{P}(k)$ is independent of k, and the corresponding Markov chain is said to be stationary.

The GCM was originally invented for global analysis of nonlinear dynamic system such as invariant sets, stable and unstable manifolds of saddle-like equilibrium states, domains of attraction and basin boundaries. These global properties of nonlinear system response are determined by the topology of the transition probability matrix $\mathbf{P}$ of the Markov chain [1–4].

When the mathematical model of a nonlinear dynamical system is known, the GCM can be created by simulating the equations of motion and using the method

J.-Q. Sun, *Random Vibration with Machine Learning Method*,
Nonlinear Physical Science, https://doi.org/10.1007/978-981-95-0812-9_12

of sampling point [2, 5–7]. Simulations start from a number of sampled initial conditions in each cell and to generate trajectories with duration of one mapping time interval. Since simulations must include with all the cells in the state space, it becomes computationally intensive for high dimensional problems where the number of cells grows exponentially with the dimension.

The one-step transition probabilities p_{ij} are estimated using the histogram on the cellular space with finite number of sampled trajectories. There exists a relationship between the number of samples and the size of the cell that minimizes the statistic error of estimation [8].

12.2 GCM from Data

The early simulation studies of the GCM method highlight the need for extensive data to build cell mappings for a reasonable operational domain of the system in the state space. The real measurements of random vibrations usually come in the form of long time histories of unknown initial conditions. The measurement data must contain rich information about the dynamics of the underlying system. However, even if a sensor network is used for a structure, for example, the number of independent output measurements can still be less than the number of degrees of freedom of the system or the dimension of the embedding state space.

Assume that a time series of a scalar output is denoted as

$$\mathbf{x} = [x_1, x_2, \dots, x_N], \tag{12.3}$$

where the subscript indicates the discrete time instant. The sample time step Δt should be sufficiently small so that the range of Nyquist frequency $(0, \frac{1}{(2\Delta t)})$ contains most important dynamics of the system. In order to construct cell mappings from the data and the one-step transition probability matrix, we must make the following assumption.

- The stochastic process of the data must be stationary.

Two questions must be answered before the cell mappings can be constructed in the state space.

1. What is the dimension of the state space in which the system resides? Another way to ask the same question is: What is the degree of freedom of the dynamic system behind the measurements?
2. From the measured random response, how do we determine the mapping time step?

The following sections present the material published in [9].

12.2.1 Determination of Embedding Dimension

Determination of the embedding dimension of the underlining system from output measurements has been studied for a long time. Several classical methods have been developed in the literature, including the multidimensional scaling (MDS) method [10], singular value decomposition (SVD) [11], principal component analysis (PCA) [12]. The embedding dimension analysis based on improved false nearest neighbors (FNN) [13] can work nicely to estimate the principal dimensions from the original data with weak noises. The method is developed based on the classical idea that some data points that are in fact far apart in the original state space may become very close to one another as nearest neighbors because the dynamics are folded in a much low-dimensional projected space. Hence, they are the false nearest neighbors in the view of real dynamic structure. The optimal embedding dimension d_m can be determined by identifying when the proportions of false nearest neighbors in data, for the first time, approach to either essentially zero or become very small with the gradual increase of possible embedding dimensions.

In the method of improved false nearest neighbors, denoted as improved FNN, the scalar time series in Eq. (12.3) can be reconstructed into a set of time-delay vectors as follows:

$$\begin{aligned} \mathbf{Y}_i^{d_m} &= (x_i, x_{i+\tau_m}, \dots, x_{i+(d_m-1)\tau_m}), \\ i &= 1, 2, \dots, N - (d_m - 1)\tau_m, \end{aligned} \tag{12.4}$$

where $\mathbf{Y}_i^{d_m}$ represents the ith reconstructed vector with the embedding dimension d_m, and τ_m denotes an index of time delay such that each component of the reconstructed vectors is independent as far as possible. In the examples below, the parameter will be given by the *C-C* method [14]. Here, the distances between each point and its nearest neighbor in dimension d_m and dimension $d_m + 1$, are measured respectively. The mean of their ratios E^{d_m} is calculated to avoid random error.

$$E^{d_m} = \frac{1}{N - d_m\tau_m} \sum_{i=1}^{N-d_m\tau_m} \frac{\left\| \mathbf{Y}_i^{d_m+1} - \mathbf{Y}_{n(i,d_m)}^{d_m+1} \right\|}{\left\| \mathbf{Y}_i^{d_m} - \mathbf{Y}_{n(i,d_m)}^{d_m} \right\|}, \tag{12.5}$$

where $\mathbf{Y}_{n(i,d_m)}^{d_m}$ represents the nearest neighbor of $\mathbf{Y}_i^{d_m}$ in the d_m-dimensional reconstructed space $\mathbb{R}^{d_m}$ in the Euclidian distance. In order to check the variation of E^{d_m} from d_m to $d_m + 1$, the $E_1^{d_m}$ is defined by

$$E_1^{d_m} = \frac{E^{d_m+1}}{E^{d_m}}. \tag{12.6}$$

In this way, the optimal embedding dimension can be determined by searching different d_m until $E_1^{d_m}$ stops changing. Then, the convergence of $E_1^{d_m}$ means that the majority of points in the data have true nearest neighbors when the embedding space has a dimension d_m. In other words, the projection of responses can be unfolded back to a multivariate state space that is representative of the original system.

When the measurements are noisy and random, $E_1^{d_m}$ may not converge with the increase of d_m. Another index, E_2, is considered to check if noise is predominant in the $(d_m + 1)^{th}$ dimension space.

$$E_2^{d_m} = \frac{\tilde{E}^{d_m+1}}{\tilde{E}^{d_m}}, \tag{12.7}$$

where

$$\tilde{E}^{d_m} = \frac{1}{N - d_m \tau_m} \sum_{i=1}^{N-d_m\tau_m} \left| x_{i+d_m\tau_m} - x_{n(i,d_m)+d_m\tau_m} \right|. \tag{12.8}$$

If the data is totally random, $E_2^{d_m}$ returns 1 for any d_m. Otherwise the value of $E_2^{d_m}$ is closely associated with parameter d_m. In this case, $E_2^{d_m}$ close to 1 indicates that the dynamics of system can be completely unfolded in a d_m-dimensional space.

The embedding dimension obtained by the method of improved false nearest neighbors does not strongly depend on how many data points are used [14]. A too small size of data will, no doubt, result in considerable error while a too large size of data will surely lead to a high computing cost. For a given dynamic system, there should be an optimal size of data.

12.2.2 Autoencoder for Embedding Dimension

A typical example of *autoencoder* is shown in Fig. 12.1. It consists of two main parts:

- **Encoder**: Compresses the input into a lower-dimensional latent space.
- **Decoder**: Reconstructs the input from the latent space back to its original form.

The goal is to minimize the difference between the input and the reconstructed output. The loss function is the mean squared error (MSE) of the prediction $\mathcal{L} = \|x - \hat{x}\|^2$.

Autoencoder has various applications such as finding an efficient representation of the data resulting dimensionality reduction or providing feature extraction. We take advantage of its ability of dimensionality reduction to find embedding dimension of the underlining dynamic system from time series of measured system outputs.

Let x_i be the input data, and $\hat{x}_i$ be the reconstructed output. We form a delayed vector of the data $\mathbf{x}_i = \{x_{i+1}, x_{i+2}, \ldots, x_{i+k}\}$ with a sufficiently large dimension k. $\hat{\mathbf{x}}_i = \{\hat{x}_{i+1}, \hat{x}_{i+2}, \ldots, \hat{x}_{i+k}\}$ is the output of the same dimension as the input. Let the number of neurons in the middle layer be m. The number of hidden layers between the input and the middle layer, i.e. the encoder, should be chosen so that the network has sufficient capacity to model the data accurately. The decoder has a symmetrical structure to the encoder with respect to the middle layer.

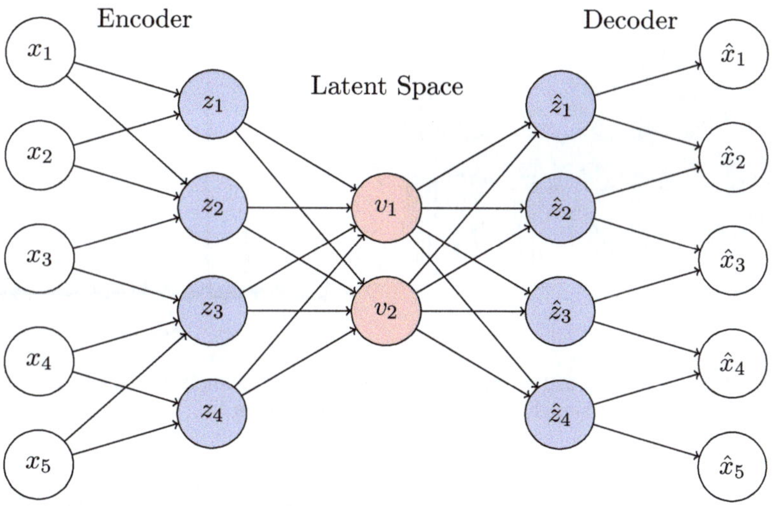

Fig. 12.1 An example of autoencoder neural networks

To find the minimum embedding dimension of the data, we start with a relatively large m and minimize the loss with respect to the trainable coefficients $\mathbf{w}$.

$$\min_{\mathbf{w}} \|\mathbf{x}_i - \hat{\mathbf{x}}_i\|^2. \tag{12.9}$$

We gradually decrease m until the loss becomes unacceptable. The smallest value of m such that the loss is still sufficiently small is a good estimate of the embedding dimension.

We can also consider the L^1 regularization in the loss function.

$$\mathcal{L}(\mathbf{w}, \lambda_j) = \|\mathbf{x}_i - \hat{\mathbf{x}}_i\|^2 + \lambda_j \sum_{j=1}^{m} |v_j|, \tag{12.10}$$

where λ_j are tuning parameters and v_j are values of the middle layer activations. In the L^1 regularization, we start with a relatively large m of the middle layer and penalize the neurons with outputs smaller than a pre-determined threshold. The algorithm to implement the L^1 regularization can be found in [15], [16], [17].

12.2.3 Determination of Mapping Time

Now, we study how to choose a proper time step for cell mappings. A stroboscopic method, similar to constructing Poincaré map, has been commonly used in the dynamic analysis of nonlinear systems. When the system is clearly periodic,

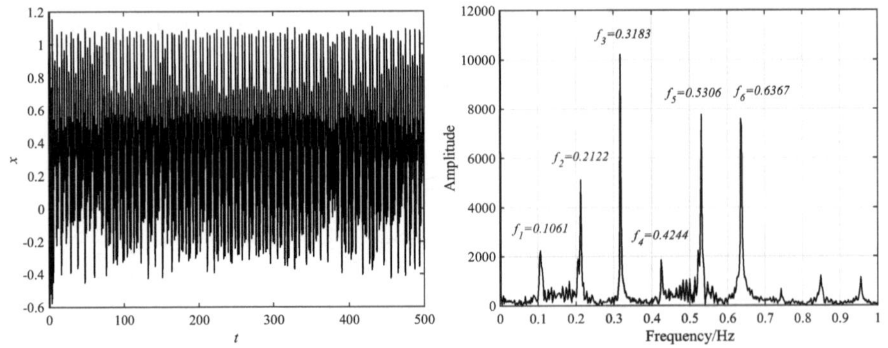

Fig. 12.2 The time-history and its spectrum diagram from data sample of Mathieu-Duffing oscillator with 1% noise. Left: Time-history. Right: Spectrum diagram

its period is used as the time step of mapping. The response of nonlinear dynamic systems subject to random excitations may contain multiple frequency components.

In order to appreciate how to pick up a proper component of frequency as the time step of mapping, a representative example of Mathieu-Duffing oscillator is considered next. The data are collected based on Monte-Carlo simulations from the governing equations by stochastic Runge-Kutta scheme with 500 Hz sampling frequency. Figure 12.2 shows one of the representative samples and its spectrum diagram, where some predominant components of frequency f_i $(i = 1, 2, \ldots, 6)$ are observed to imply the periodicities exhibited by the system.

Assume that the principal dimension has been determined. A Poincaré section on the $x - \dot{x}$ plane is constructed in Fig. 12.3 with different time intervals $1/f_i$. It is found that when the choosing mapping step τ is the integer multiple of the driving period T, the patterns of the return points on the Poincaré section remain close to the real ones. The mapping step is proposed to be determined by the following process:

1. Determine the periodicities of the collected data by discrete Fourier transform.
2. Construct Poincaré sections in a predetermined d_m-dimensional state space with the mapping steps determined from different dominant frequencies. Find same patterns of the return points and record the corresponding frequencies f_i.
3. Choose any one of f_i as a base-frequency f_0 and let $n = f_0/f_i$. Check if other f_i satisfies that when n is an integer the patterns of Poincaré sections recur and when n is a fraction the patterns are doubling. If not, replace f_0 with another f_i and repeat step 3.

For the Mathieu-Duffing oscillator, the mapping step is determined as $\tau = 1/f_3 = 3.1417s$. On the other hand, when there are no obvious dominant frequencies in the spectrum, or when the system is known to be autonomous, the selection of mapping step can be arbitrary and should be related to the size of cells and the number of data points N [8].

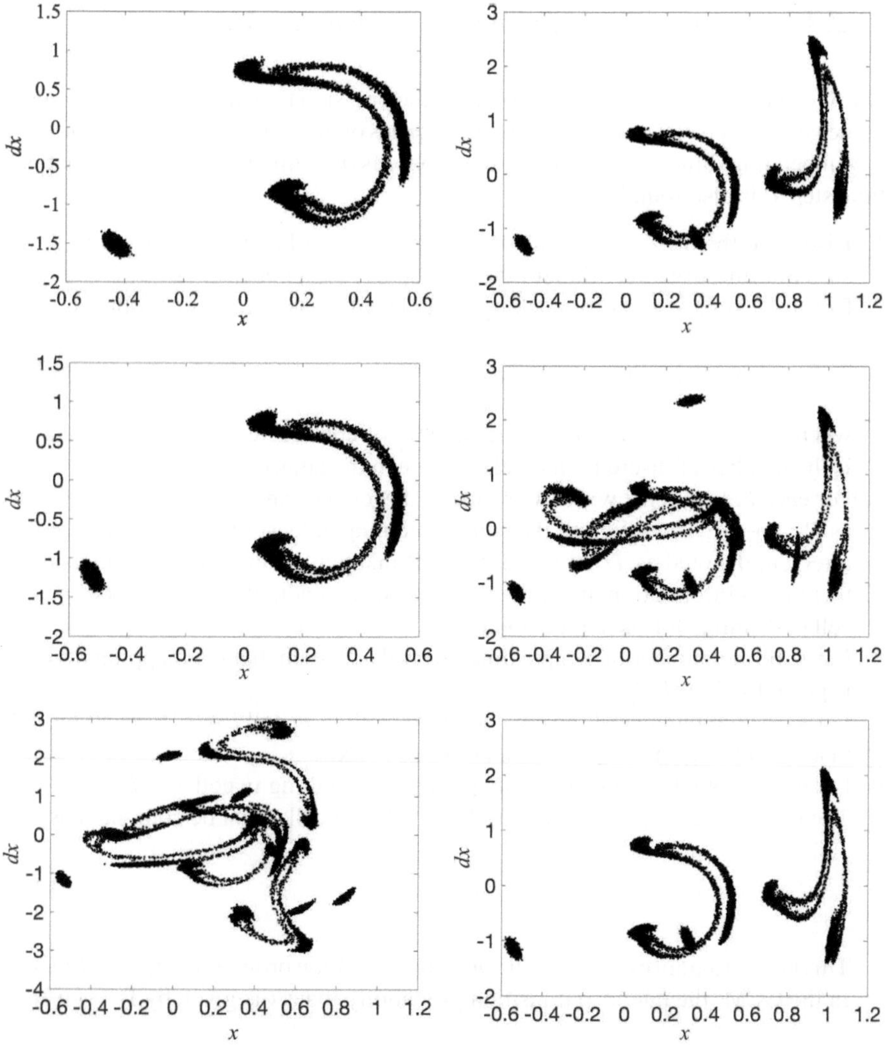

Fig. 12.3 The return points of the response orbits on the Poincaré section with different time intervals $1/f$ with different N for the case of $D_m = 3$. Top row: $f = 0.1061$ and $f = 0.2122$. Middle row: $f = 0.3183$ and $f = 0.4244$. Bottom row: $f = 0.5306$ and $f = 0.6367$

12.2.4 Construction of Transition Probability Matrix

An algorithm is presented next to compute the one-step transition probability matrix of the generalized cell mapping from a long series of measurements. It is assumed that the embedding dimension d_m of the state space is determined and a proper mapping time step T is also found.

1. Discretize the d_m-dimensional state space into a collection of cells with proper size. Starting from the origin of the state space, we label each cell by an integer i.
2. Construct state vectors by stacking the delayed measurements as follows.

$$\mathbf{x}_k = \{x_k, x_{k+\Delta k}, x_{k+2\Delta k}, \ldots, x_{k+(d_m-1)\Delta k}\}^T \in \mathbb{R}^{d_m \times 1} \tag{12.11}$$

   where $k = 1, 2, \ldots, N_s$, N_s is the number of constructed state vectors, $\Delta k \geq 1$ is the number of discrete time delay between the adjacent state variables.
3. For each k, find cell i where $\mathbf{x}_k$ resides. This creates an integer mapping denoted as $P : k \mapsto i$. The set of all the cells containing at least one data point $\mathbf{x}_k$ covers a region, denoted as $\mathcal{D}$, in which the cell mappings can be defined. Outside $\mathcal{D}$, there is no information about the system. Let $\mathcal{I}$ denote the set of indices i of the cell containing at least a data point.
4. Let K be the number of sample time intervals Δt such that the mapping time step is given by $T = K\Delta t$.
5. Let n_{k_i} be the number of the data points $\mathbf{x}_{k_i}$ falling in cell $i \in \mathcal{I}$. Find all the cells that contain the one-step images, i.e. $\mathbf{x}_{k_i+K}$, of $\mathbf{x}_{k_i}$.
6. Let $n_{i,j}$ denote the number of data points $\mathbf{x}_{k_i+K}$ falling in cell $j \in \mathcal{I}$.
7. The one step transition probability from cell i to cell j can be computed as

$$p_{ij} = \frac{n_{i,j}}{n_{k_i}}. \tag{12.12}$$

   This is a histogram estimate of the one-step transition probability density function. In the GCM, the one-step transition probability matrix is given by $\mathbf{P} = \{p_{ij}\}$.

12.3 Example of Mathieu-Duffing Oscillator

Data-driven global analysis for a Mathieu-Duffing oscillator is carried out. An additive Gaussian white noise with intensity σ excites the system.

Consider a two-dimensional Mathieu-Duffing oscillator, which is governed by the following equation of motion

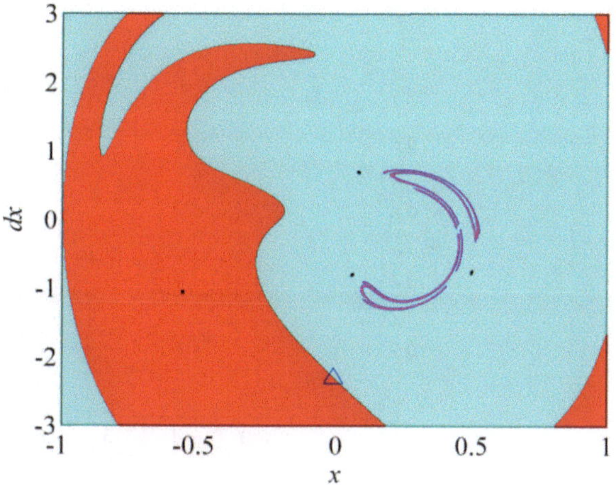

Fig. 12.4 Global structure of deterministic Mathieu-Duffing oscillator, where black "+" stands for period-1 attractor, black "*" for period-3 attractor, magenta "·" for a chaotic saddle, green "Δ" for a saddle point on basin boundary

$$\frac{d^2x}{dt^2} + 25x^3 + 0.173\frac{dx}{dt} + [2.62 - 1.7328(1 - \cos 2t)]\,x$$
$$= 3.496(1 - \cos 2t) + \sigma\omega(t), \tag{12.13}$$

where $\omega(t)$ is a unit white noise process. For the deterministic case ($\sigma = 0$), the global behaviors of system in the state space include a period-1 attractor (+), a period-3 attractor (*), a chaotic saddle ($\cdot$) as well as a saddle point (Δ) on the basin boundary (red solid curves), which are shown in Fig. 12.4.

We simulate Eq. (12.13) to generate data with the noise intensity $\sigma = 1\%$ from only one initial condition located in the basin of period-3. The trajectory is 2000 s long with a sampling frequency of 500 Hz. The time-history and the frequency components are shown in Fig. 12.2. The mapping time step is chosen to be $\tau = 1/f_3 = 3.1416s$. In this example, the last two thousand data points are extracted to estimate the embedding dimension. From Fig. 12.5, E_1 and E_2 values approximately equal to 1 when $d_m = 2$. Therefore, it is apparent that the embedding dimension of the system is 2.

Thus, the state space is two-dimensional. The domain $\mathcal{D}$ is chosen to be $\{-1 < S_k < 1, -1 < S_{k+1} < 1\}$. $\mathcal{D}$ is discretized into 200×200 cells. The depth-first search (DFS) algorithm based on the graph theory is executed to analyze strongly connected digraphs for finding the stable and unstable invariants.

Consider the 3% noise intensity. Random initial conditions uniformly distributed in the state space are taken to collect multiple sampling data. Here, the number of initial points is set as 20. Several samples are drawn in Fig. 12.6. In this case, the proposed data-driven GCM method can work well to capture the unstable chaotic saddle as well as the two coexistent periodic attractors in Fig. 12.7.

When the noise intensity is 25%, a noise-induced dangerous bifurcation is observed as reported in [18] that the collision between the stochastic period-1 attractor with a stochastic saddle on the basin boundary causes the blue-sky disappearance of the stochastic period-1 attractor. Then, the system responses almost surely exit the

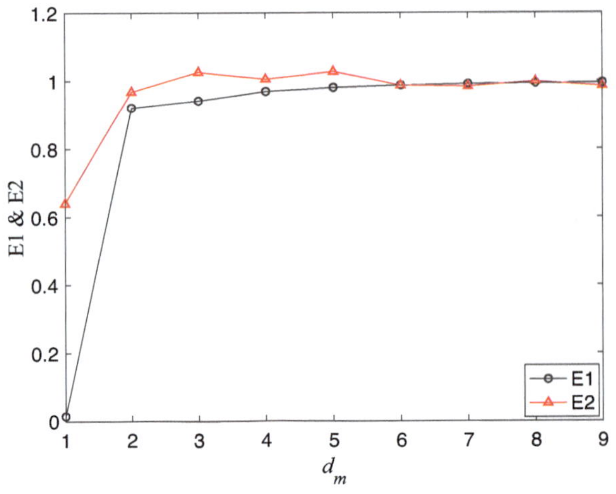

Fig. 12.5 The values of E_1 and E_2 for data from Mathieu-Duffing oscillator

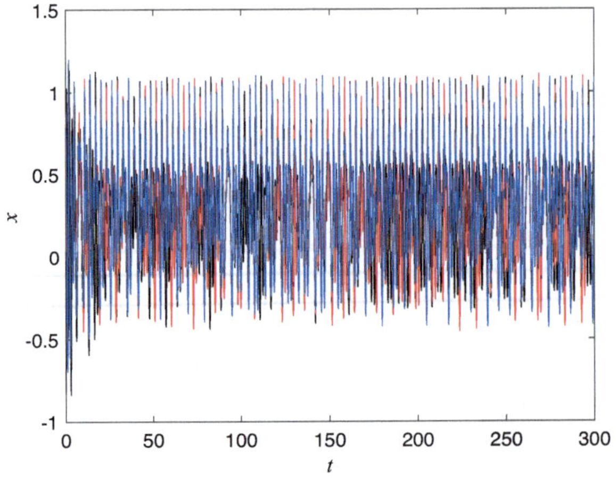

Fig. 12.6 The time-histories from data samples of Mathieu-Duffing oscillator with 3% noise

basin of period-1 attractor and enter the region around the period-3 attractor through the stochastic saddle on the basin boundary. Figure 12.8 depicts the evolution of trajectories. Although the creation of transition probability matrix may benefit from that the stochastic responses spread over a much wider region and visit more cells due to the nonlinear phenomenon of noise-induced response transitions, whereas the global structures of the system may become such explosive that some details of dynamical properties are obscured because of such a strong noise, as shown in Fig. 12.9.

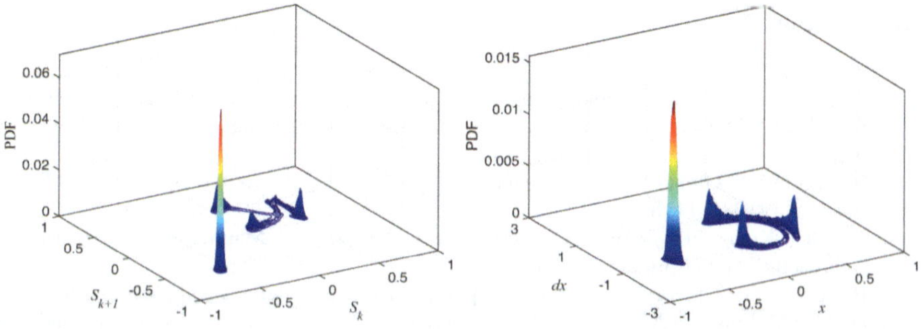

Fig. 12.7 The global dynamics and its probability distribution from data of Mathieu-Duffing oscillator with 20 initial conditions and 3% noise in different spaces with 200 × 200 cell resolutions. Left: Mapping space $S_k - S_{k+1}$. Right: State space $x - \dot{x}$

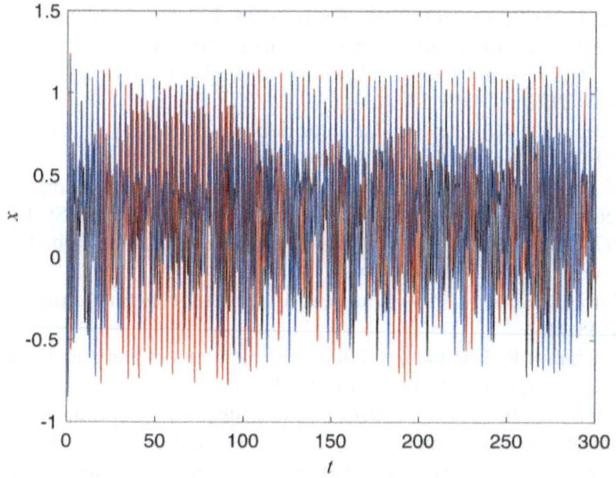

Fig. 12.8 The time-histories from data samples of Mathieu-Duffing oscillator with 25% noise

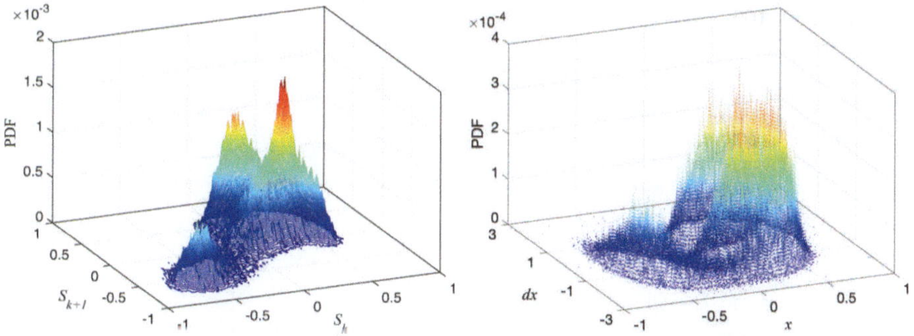

Fig. 12.9 The global dynamics and its probability distribution from data of Mathieu-Duffing oscillator with 20 initial conditions and 25% noise in different spaces with 200 × 200 cell resolutions. Left: Mapping space $S_k - S_{k+1}$. Right State space $x - \dot{x}$

References

1. Chung KL (1967) Markov chains with stationary transition probabilities. Springer, New York
2. Hsu CS (1982) A probabilistic theory of nonlinear dynamical systems based on the cell state space concept. J Appl Mech 49:895–902
3. Hsu CS (1995) Global analysis of dynamical-systems using posets and digraphs. Int J Bifurc Chaos 5(4):1085–1118
4. Sun JQ, Xiong FR, Schütze O, Hernández C (2019) Cell Mapp Methods-Algorithmic Approaches Appl. Springer, Berlin
5. Hsu CS (1980) A theory of cell-to-cell mapping dynamical systems. J Appl Mech 47:931–939
6. Dellnitz M, Junge O (2002) Set oriented numerical methods for dynamical systems. Handbook of dynamical systems 2:221–264
7. Zou H, Xu J (2009) Improved generalized cell mapping for global analysis of dynamical systems. Sci China Ser E: Technol Sci 52(3):787–800
8. Sun JQ, Hsu CS (1988) A statistical study of generalized cell mapping. J Appl Mech 55:694–701
9. Li Z, Jiang J, Hong L, Sun JQ (2019) On the data-driven generalized cell mapping method. Int J Bifurc Chaos 29(14):1950,204. https://doi.org/10.1142/S0218127419502043
10. Machado J (2012) Multidimensional scaling analysis of fractional systems. Comput Math Appl 64:2966–2972
11. Klema V, Laub A (1980) The singular value decomposition: its computation and some applications. IEEE Trans Autom Control 25:164–176
12. Vanhatalo E, Kulahci M, Bergquist B (2017) On the structure of dynamic principal component analysis used in statistical process monitoring. Chemom Intell Lab Syst 167:1–11
13. Cao L (1997) Practical method for determining the minimum embedding dimension of a scalar time series. Phys D 110:43–50
14. Kim H, Eykholt R, Salas J (1999) Nonlinear dynamics, delay times, and embedding windows. Phys D 127:48–60
15. Hastie T, Tibshirani R, Friedman J (2009) The elements of statistical learning: data mining, inference, and prediction. Springer series in statistics
16. Brunton SL, Proctor JL, Kutz JN (2016) Discovering governing equations from data by sparse identification of nonlinear dynamical systems. Proc Natl Acad Sci 113(15):3932
17. Lai Z, Nagarajaiah S (2019) Sparse structural system identification method for nonlinear dynamic systems with hysteresis/inelastic behavior. Mech Syst Signal Process 117:813–842
18. Li Z, Jiang J, Hong L (2015) Transient behaviors in noise-induced bifurcations captured by generalized cell mapping method with an evolving probabilistic vector. Int J Bifurc Chaos 25(1550):109

Chapter 13
New Application of Simulation

Discussions presented in Chap. 3 deal with how to simulate random random variable, stochastic processes and responses of random vibration problems. The simulated responses are often used as benchmarks to validate approximate solutions obtained by analytical or numerical methods. In this sense, we can say that Monte Carlo simulation plays a central role in research.

Recently, Monte Carlo simulation is given an assistive role. That is, simulation results are used to assist other analytical and numerical methods by providing important intermediate and yet critical results. In this chapter, two examples are presented to illustrate how simulation can be used to assist effective implementation of other methods. The first example makes use of Monte Carlo simulation as an assistive tool for optimal selection of Gaussian functions of the RBFNN method for high dimensional problems in random vibration. The second example presents a scheme to reduce the order of or decouple the multiple degree of freedom system with the help of limited simulation results.

13.1 Design of RBFNN

The radial basis function neural networks (RBFNN) with Gaussian activation functions have been intensively applied to obtain solution of the FPK equations for smooth [1–3] and non-smooth systems [4, 5] and the first-passage reliability problem [6].

In this section, we show how to use simulations to help design optimal RBFNN for a given nonlinear stochastic system. Specifically, we present a strategy to find the optimal number and locations of the means of Gaussian neuron by taking advantages of rough probability distribution of the steady-state response obtained by Monte Carlo simulations. An index is introduced to assess the contribution of a Gaussian neuron to the total solution in order to develop a RBFNN with the best performance and minimum computational effort. The material is taken from [7].

J.-Q. Sun, *Random Vibration with Machine Learning Method*,
Nonlinear Physical Science, https://doi.org/10.1007/978-981-95-0812-9_13

Consider an N-dimensional nonlinear dynamical system under Gaussian white noise excitations.

$$\frac{dX_i}{dt} = g_i(\mathbf{X}) + \sum_{j=1}^{m} h_{ij}(\mathbf{X})W_j(t), \quad 1 \le i \le N, \tag{13.1}$$

where $g_j(\mathbf{X})$ and $h_{jk}(\mathbf{X})$ are nonlinear functions of $\mathbf{X}$. $W_j(t)$ are independent Gaussian white noises with zero mean and the correlation matrix $\mathrm{E}[W_j(t)W_j(t-\tau)] = 2D_j\delta(\tau)$ $(j = 1, 2, \ldots, m)$. It is assumed that when $t \to \infty$, a stationary PDF $p_{\mathbf{X}}(\mathbf{x})$ of the response exists and satisfies the following reduced FPK Eq. (13.2),

$$\begin{aligned} \mathcal{L}_{FPK}[p_{\mathbf{X}}(\mathbf{x})] \equiv & -\sum_{i=1}^{N} \frac{\partial}{\partial x_i}[m_i(\mathbf{x})p_{\mathbf{X}}(\mathbf{x})] \\ & + \sum_{i=1}^{N}\sum_{j=1}^{N} \frac{\partial^2}{\partial x_i \partial x_j}\left[\frac{b_{ij}(\mathbf{x})}{2} p_{\mathbf{X}}(\mathbf{x})\right] = 0, \end{aligned} \tag{13.2}$$

where $\mathbf{x} = \{x_1, x_2, \ldots, x_N\} \in \mathbb{R}^N$ is the position vector in the state space. $\mathcal{L}_{FPK}[\cdot]$ denotes for the differential operator of the FPK equation. $m_i(\mathbf{x})$ and $b_{ij}(\mathbf{x})$ are known as the drift and diffusion terms defined as,

$$\begin{aligned} m_i(\mathbf{x}) &= g_i(\mathbf{x}) + \sum_{k=1}^{N}\sum_{j=1}^{m}\left[D_j h_{kj}(\mathbf{x})\frac{\partial h_{ij}(\mathbf{x})}{\partial x_k}\right], \\ b_{ij}(\mathbf{x}) &= \sum_{k=1}^{m} 2D_k h_{ik}(\mathbf{x})h_{jk}(\mathbf{x}). \end{aligned} \tag{13.3}$$

The PDF satisfies the following normalization condition.

$$\int_{\mathbb{R}^N} p_{\mathbf{X}}(\mathbf{x})d\mathbf{x} = 1. \tag{13.4}$$

The PDF solution of the reduced FPK Eq. (13.2) is expressed in the form of RBFNNs as,

$$\bar{p}_{\mathbf{X}}(\mathbf{x}, \mathbf{w}) = \sum_{j=1}^{N_G} w_j g_j(\mathbf{x}), \tag{13.5}$$

where N_G is the number of Gaussian neurons with the mean $\boldsymbol{\mu}_j$ and covariance matrix $\boldsymbol{\Sigma}_j$ defined as

$$g_j(\mathbf{x}) \equiv g(\mathbf{x}, \boldsymbol{\mu}_j, \boldsymbol{\Sigma}_j) \qquad (13.6)$$
$$= \frac{1}{\sqrt{(2\pi)^N \det(\boldsymbol{\Sigma}_j)}} \exp\left[-\frac{1}{2}(\mathbf{x}-\boldsymbol{\mu}_j)^T \boldsymbol{\Sigma}_j^{-1}(\mathbf{x}-\boldsymbol{\mu}_j)\right].$$

The RBFNN PDF solution must satisfy the normalization condition in Eq. (13.4). Note that the Gaussian function $g_j(\mathbf{x})$ defined in Eq. (13.6) is normalized such that

$$\int_{\mathbb{R}^N} g_j(\mathbf{x})d\mathbf{x} = 1, \quad \text{for all } j. \qquad (13.7)$$

As a result, the normalization condition in Eq. (13.4) leads to,

$$\sum_{j=1}^{N_G} w_j = 1. \qquad (13.8)$$

The error of the FPK equation incurred by the RBFNN solution is defined as

$$e(\mathbf{x}, \mathbf{w}) = -\sum_{i=1}^{N} \frac{\partial}{\partial x_i}[m_i(\mathbf{x})\bar{p}_{\mathbf{X}}(\mathbf{x}, \mathbf{w})] \qquad (13.9)$$
$$+ \sum_{i=1}^{N}\sum_{j=1}^{N} \frac{\partial^2}{\partial x_i \partial x_j}\left[\frac{b_{ij}(\mathbf{x})}{2}\bar{p}_{\mathbf{X}}(\mathbf{x}, \mathbf{w})\right].$$

13.1.1 Neuron Performance Index

We run Monte Carlo simulations of the system starting from initial positions randomly distributed in a region D_G of the state space in order to obtain a rough distribution of the response. Let N_r be the number of random points $\mathbf{x}_i$ representing the response in the steady state. Let $\mathbf{X}_r = \{\mathbf{x}_1, \ldots, \mathbf{x}_{N_r}\}$ denote the set of these random points. We define an index $I(j)$ of the jth neuron to evaluate its contribution within the set $\mathbf{X}_r$ as,

$$I(j) = \frac{1}{Q}\sum_{i=1}^{N_r} g(\mathbf{x}_i, \boldsymbol{\mu}_j, \boldsymbol{\sigma}), \ \mathbf{x}_i \in \mathbf{X}_r, \qquad (13.10)$$

where Q is a normalization coefficient defined as

$$Q = \sum_{j=1}^{N_G}\sum_{i=1}^{N_r} g(\mathbf{x}_i, \boldsymbol{\mu}_j, \boldsymbol{\sigma}). \qquad (13.11)$$

Hence, we have

$$\sum_{j=1}^{N_G} I(j) = 1. \tag{13.12}$$

The Gaussian neuron possesses the following property:

$$\int_{\mathbb{R}^N} g(\mathbf{x}_i, \boldsymbol{\mu}, \boldsymbol{\sigma}) d\boldsymbol{\mu} = \int_{\mathbb{R}^N} g(\mathbf{x}, \boldsymbol{\mu}_j, \boldsymbol{\sigma}) d\mathbf{x} = 1, \text{ for all } i \text{ and } j. \tag{13.13}$$

Let d_v denote the differential volume element of the grid for the mean $\boldsymbol{\mu}$. We have

$$d_v = \prod_{k=1}^{N} h_k. \tag{13.14}$$

An approximate expression for Q can be obtained as

$$\begin{aligned} Q &= \sum_{j=1}^{N_G} \sum_{i=1}^{N_r} g(\mathbf{x}_i, \boldsymbol{\mu}_j, \boldsymbol{\sigma}) \\ &= d_v^{-1} \sum_{i=1}^{N_r} \left[\sum_{j=1}^{N_G} g(\mathbf{x}_i, \boldsymbol{\mu}_j, \boldsymbol{\sigma}) d_v \right] \\ &\approx d_v^{-1} \sum_{i=1}^{N_r} \int_{\mathbb{R}^N} g(\mathbf{x}_i, \boldsymbol{\mu}, \boldsymbol{\sigma}) d\boldsymbol{\mu} \quad \text{as } N_G \to \infty \\ &= d_v^{-1} N_r. \end{aligned} \tag{13.15}$$

13.1.2 Statistical Properties of $I(j)$ and $w(j)$

We discretize the state space so that the mean vector $\boldsymbol{\mu}_j$ is at the jth grid and is the center of a box with the size d_v defined in Eq. (13.14). Let n_j denote the number of samples falling into the box centered at the jth grid. $p_r(\boldsymbol{\mu}_j) = n_j/N_r$ is a histogram estimate of the probability of random points $\mathbf{X}_r$ found in the box. By definition, $\sum_{j=1}^{N_G} n_j = N_r$.

The index $I(j)$ can be computed by collecting points in the boxes of the discretized state space. Making use of the result in Eq. (13.15) and $p_r(\boldsymbol{\mu}_j)$, we obtain,

$$I(j) \approx d_v \sum_{i=1}^{N_G} p_r(\boldsymbol{\mu}_i) g(\boldsymbol{\mu}_i, \boldsymbol{\mu}_j, \boldsymbol{\sigma}). \tag{13.16}$$

Hence, the mean of the index is given by

$$\mathrm{E}[I(j)] = d_v \sum_{i=1}^{N_G} \mathrm{E}[p_r(\boldsymbol{\mu}_i)] g(\boldsymbol{\mu}_i, \boldsymbol{\mu}_j, \boldsymbol{\sigma}). \tag{13.17}$$

When the standard deviations of all the Gaussian neurons are sufficiently small such that approximately we have

$$g(\boldsymbol{\mu}_i, \boldsymbol{\mu}_j, \boldsymbol{\sigma}) = (2\pi)^{-N/2} d_v^{-1} \delta_{ij}, \tag{13.18}$$

where δ_{ij} is the Kronecker delta function. Hence, we have

$$\mathrm{E}[I(j)] = (2\pi)^{-N/2} \mathrm{E}[p_r(\boldsymbol{\mu}_j)]. \tag{13.19}$$

This suggests that $\mathrm{E}[I(j)]$ is approximately proportional to the probability of the system falling in the box centered at the jth grid. $I(j)$ is essentially a smoothed version of $p_r(\boldsymbol{\mu}_j)$ weighted by the Gaussian function values at their mean locations.

Let $\mathrm{var}[p_r(\boldsymbol{\mu}_j)]$ denote the variance of $p_r(\boldsymbol{\mu}_j)$. Under the same assumption as in Eq. (13.18), the variance of $I(j)$ can be obtained as,

$$\mathrm{var}[I(j)] = \mathrm{var}[p_r(\boldsymbol{\mu}_j)](2\pi)^{-N}. \tag{13.20}$$

When $N_r \to \infty$, $p_r(\boldsymbol{\mu}_j)$ converges to the true PDF solution of the system at $\mathbf{x} = \boldsymbol{\mu}_j$. According to the solution in Eq. (13.5), we must have

$$p_r(\boldsymbol{\mu}_j) = \sum_{i=1}^{N_G} w_i g(\boldsymbol{\mu}_j, \boldsymbol{\mu}_i, \boldsymbol{\sigma}). \tag{13.21}$$

Make use of Eq. (13.18). By definition, we have the index as

$$\begin{aligned}
I(j) &= d_v \sum_{i=1}^{N_G} p_r(\boldsymbol{\mu}_i) g(\boldsymbol{\mu}_i, \boldsymbol{\mu}_j, \boldsymbol{\sigma}) \\
&= \sum_{i=1}^{N_G} \sum_{k=1}^{N_G} w_k g(\boldsymbol{\mu}_i, \boldsymbol{\mu}_k, \boldsymbol{\sigma}) g(\boldsymbol{\mu}_i, \boldsymbol{\mu}_j, \boldsymbol{\sigma}) d_v \\
&= \sum_{i=1}^{N_G} \sum_{k=1}^{N_G} w_k (2\pi)^{-N/2} d_v^{-1} \delta_{ik} (2\pi)^{-N/2} d_v^{-1} \delta_{ij} d_v \\
&= w_j (2\pi)^{-N} d_v^{-1}.
\end{aligned} \tag{13.22}$$

Hence, $I(j)$ and $w(j)$ are proportional to each other. This also means that the value of $I(j)$ is an indicator of the contribution of the jth neuron.

13.1.3 Optimal Selection Strategy

Recall that $\sum_{j=1}^{N_G} I(j) = 1$ when all the grid points in D_G are taken as the mean of the Gaussian neuron. Let $\mathcal{G}_\epsilon$ denote the set of Gaussian neurons with top rank performance indices such that

$$\sum_{j \in \mathcal{G}_\epsilon} I(j) = \epsilon, \tag{13.23}$$

where $0 < \epsilon \leq 1$ is a pre-selected threshold. In the following examples, we set $\epsilon = 0.999$. ϵ is a hyper-parameter of the proposed selection method. The threshold 0.999 was experimentally found to balance the accuracy and efficiency well. Let N_G^s be the number of neurons in $\mathcal{G}_\epsilon$. $1 < N_G^s \leq N_G$.

To evaluate the accuracy of the RBFNN solution to the FPK equation, two different errors are introduced. The first is the root mean square error of the FPK equation given by:

$$J_{rms}(\mathbf{w}) = \sqrt{\int_{\mathbb{R}^N} e(\mathbf{x}, \mathbf{w})^2 d\mathbf{x}}, \tag{13.24}$$

where $e(\mathbf{x}, \mathbf{w})$ is the local error defined in Eq. (13.9). This error is used to judge the quality of the RBFNN solution to satisfy the FPK equation.

The second is the error of the RBFNN solution in comparison to available analytical or simulation results,

$$\Delta p = \sqrt{\int_{\mathbb{R}^N} [p_{MC}(\mathbf{x}) - \bar{p}_{\mathbf{X}}(\mathbf{x}, \mathbf{w})]^2 d\mathbf{x}}, \tag{13.25}$$

where $p_{MC}(\mathbf{x})$ denotes the PDF computed from Monte Carlo simulations. The second error measures the accuracy of the solution directly. These two error terms are related, but different.

We replace w_j in Eq. (13.24) with $I(j)$ to compute an RMS error, denoted as e_r, of the FPK equation and study the convergence of e_r in order to find an optimal selection of Gaussian neurons without computing the RBFNN solution of the FPK equation with the help of w_j.

$$e_r(\mathbf{I}) = \sqrt{\int_{\mathbb{R}^N} e(\mathbf{x}, \mathbf{I})^2 d\mathbf{x}}, \tag{13.26}$$

where $e(\mathbf{x}, \mathbf{I})$ is $e(\mathbf{x}, \mathbf{w})$ when $\mathbf{w} \to \mathbf{I}$, and $\mathbf{I} = [I(j)]$ is the vector of the performance index computed from rough simulations, not from the FPK equation. Hence, $e_r(\mathbf{I})$ is a rough estimate of $J_{rms}(\mathbf{w})$ without solving the FPK equation. It is used to judge the quality of the selected number and location of Gaussian neurons.

We apply an iterative algorithm, which generates more random points in each iteration until the error e_r does not change much. Let $N_r(1)$ denote the initial number of random points. We calculate the error $e_r(1)$ and the number of selected neurons $N_G^s(1)$ by making use of Eq. (13.23). In subsequent iterations, we take $N_r(k) = 2N_r(k-1)$ and calculate $N_G^s(k)$ and $e_r(k)$. The iteration will stop when $\frac{e_r(k)-e_r(k-1)}{e_r(k-1)}\% < \eta$. η is set to 10%. The neurons selected according to Eq. (13.23) are used to solve the FPK equation.

13.1.4 Van der Pol System

We take the Van der Pol system as an example to illustrate the neuron selection process.

$$\begin{aligned}\frac{dX_1}{dt} &= X_2,\\ \frac{dX_2}{dt} &= -\beta(X_1^2-1)X_2 - X_1 + W_1(t),\end{aligned} \tag{13.27}$$

where $\beta > 0$ is a system parameter. $W_1(t)$ denotes the Gaussian white noise excitation with intensity $2D_1$. We choose $\beta = 1.5$ and $D_1 = 0.25$. We should point out that this system is highly nonlinear and is known to be a challenging case to test the effectiveness of methods for analytical and numerical solutions.

The Gaussian neurons are ranked by their values of the index $I(j)$ using the simulation data of the system. The top ranked neurons are selected for the RBFNN solution.

The computational domain of the RBFNN method in the state space is $D_G = [-4, 4] \times [-7, 7]$. The number of sampling points is $N_s = 121 \times 121 = 14641$. The number of candidate neurons in D_G to be selected is $N_G = 101 \times 101 = 10201$. This number is the pre-set upper bound of the Gaussian neurons to be included in the solution. We use all N_G neurons to calculate the RBFNN PDF solution of the FPK equation. The accuracy of the RBFNN PDF solution can be assessed by the RMS error defined in Eq. (13.24) and by Monte Carlo simulations.

Monte Carlo simulations of the Van der Pol system are carried out for $T = 100\pi$ s. 10^7 trajectories starting from the origin $(0, 0)$ are obtained. The integration time step is $\tau = \pi/50$. The CPU time for simulation is 3200 s. We should note that such extensive simulations are intended to provide a basis to validate the RBFNN solution for this one-degree-of-freedom system only and are not used in the optimal neuron selection process. Furthermore, it would be far more computationally intensive to generate such extensive simulations for high dimensional systems.

For the RBFNN solution with 10201 neurons, the RMS error J_{rms} of the FPK equation is 3.87×10^{-4}, and the RMS error Δp of the PDF solution of the FPK equation compared with Monte Carlo simulations is 3.5×10^{-4}. The CPU time of the RBFNN solution is 54 s. This is a fairly accurate result and is used as a baseline

Table 13.1 The iteration process for finding optimal number of Gaussian neurons for the Van der Pol system

Iteration no. k	1	2	3	4	5	6
$N_r(k)$	800	1600	3200	6400	12800	25600
$N_G^s(k)$	2236	2473	2658	2743	2792	2812
$e_r(k)$	1.3008	0.98879	0.70488	0.57891	0.39857	0.35961
$\frac{\lvert e_r(k)-e_r(k-1)\rvert}{e_r(k)}\%$		23.99%	28.71%	17.87%	31.15%	9.77%

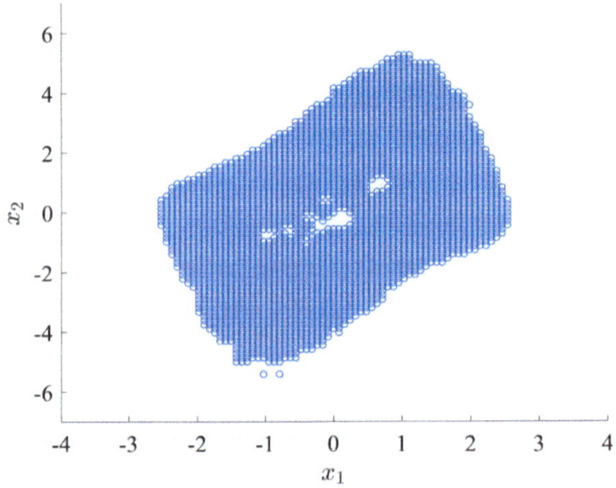

Fig. 13.1 Optimally selected mean locations of 2812 Gaussian neurons

to investigate the improvement of the RBFNN solution with the proposed neuron selection strategy.

Table 13.1 summarizes the iteration process to select Gaussian neurons. At the sixth step of iteration, the relative change of the error $e_r(k)$ is less than the set goal of 10%. The iteration process takes 10.3 s. The selection algorithm leads to 2812 Gaussian neurons with top rankings of the performance index $I(j)$. The distribution of the selected neurons is shown in Fig. 13.1. The RBFNN solution with 2812 selected neurons takes 6.4 s to obtain the network coefficients for the solution of the FPK equation. To present another fair comparison, we select $N_G = 53 \times 53 = 2809$ Gaussian neurons uniformly distributed in the same domain. We summarize the numerical results in Table 13.2. The numbers in the table clearly indicate that the RBFNN solution of the FPK equation with optimally selected neurons is most accurate and is quite efficient as well.

Table 13.2 Comparison of RMS errors of the RBFNN solutions for the Van der Pol system

N_G	Distribution	FPK RMS J_{rms}	MC RMS Δp	CPU (s)
10201	Uniform	3.87×10^{-4}	3.5×10^{-4}	54
2809	Uniform	7.2×10^{-2}	1.38×10^{-1}	6.3
2812	Optimal	2.5×10^{-3}	4.8×10^{-3}	16.7

13.1.5 Lorenz System

We consider a 3D Lorenz system governed by

$$\begin{aligned} \frac{dY_1}{dt} &= \sigma(Y_2 - Y_1), \\ \frac{dY_2}{dt} &= \rho Y_1 - Y_2 - Y_1 Y_3, \\ \frac{dY_3}{dt} &= Y_1 Y_2 - \beta Y_3, \end{aligned} \tag{13.28}$$

where the constants σ, ρ, and β are related to the Rayleigh number, the Prandtl number, and a geometrical factor. Consider a coordinate transformation defined by $X_1 = \frac{Y_1}{20}$, $X_2 = \frac{Y_2}{20}$, and $X_3 = \frac{Y_3}{20}$ and introduce additive Gaussian white noises to all the states.

$$\begin{aligned} \frac{dX_1}{dt} &= \sigma(X_2 - X_1) + W_1(t), \\ \frac{dX_2}{dt} &= \rho X_1 - X_2 - 20X_1X_3 + W_2(t), \\ \frac{dX_3}{dt} &= 20X_1X_2 - \beta X_3 + W_3(t), \end{aligned} \tag{13.29}$$

where $W_i(t)$ denote the uncorrelated Gaussian white noise excitations with intensity $2D_i$. It is well known that the deterministic Lorenz system has one chaotic solution and two fixed points.

We take $\sigma = 10$, $\rho = 18$, $\beta = \frac{8}{3}$ and $D_1 = D_2 = D_3 = 0.01$. The reduced FPK equation of the system has the following drift and diffusion terms,

$$\begin{aligned} m_1 &= \sigma(x_2 - x_1), \\ m_2 &= \rho x_1 - x_2 - 20x_1x_3, \\ m_3 &= 20x_1x_2 - \beta x_3, \\ b_{11} &= 2D_1,\ b_{22} = 2D_2,\ b_{33} = 2D_3. \end{aligned} \tag{13.30}$$

Table 13.3 The iteration process for finding optimal number and location of Gaussian neurons for the Lorenz system

Iteration no. k	1	2	3	4	5	6
$N_r(k)$	3200	6400	12800	25600	51200	102400
$N_G^S(k)$	55324	61773	66889	71049	74665	80889
$e_r(k)$	41.456	28.901	23.004	19.444	17.434	16.473
$\frac{\lvert e_r(k)-e_r(k-1)\rvert}{e_r(k)}\%$		30.29%	20.4%	15.48%	10.34%	5.51%

The domain of interest in the state space is $D_G = [-1, 1] \times [-1, 1] \times [-0.25, 1.75]$ and is discretized into $N_G = 100 \times 100 \times 100 = 10^6$ grids. We take $N_s = 105 \times 105 \times 105$ uniformly sampled points.

The iterative neuron selection algorithm is applied. The iterative process is summarized in Table 13.3. Finally, the total number of random points used in the selection is $N_r = 102400$. The number of selected neurons is $N_G^S = 80889$.

The RBFNN solution of the FPK equation with 80889 optimally selected neurons is compared with Monte Carlo simulations as shown in Figs. 13.2, 13.3 and 13.4. The memory requirement of the 80889×80889 matrix is about 0.64% of the memory needed for a $10^6 \times 10^6$ matrix. The CPU time of the RBFNN method is 1756 s, including 616 s for the iteration process, and 1140 s for the PDF solution. In comparison, Monte Carlo simulations with 10^8 samples take 77666 s. The integration time is 50π and the integration time step is 0.01π. The RMS error of the RBFNN solution of the FPK equation compared to simulations is 0.0067 and the RMS error of the FPK equation 9.28×10^{-2}.

13.2 Order Reduction

The work in [8] presents a method of model reduction to derive governing equation of the marginal PDF for a small number of state variables of special interest. The simulation is used to assist the derivation of the equation for the governing equation. This section presents a brief summary of the study reported in the reference.

13.2.1 Equation for Marginal PDF

We partition the state vector into two parts $\mathbf{x} = \{\mathbf{x}_1, \mathbf{x}_2\}$. The dimension of $\mathbf{x}_1$ is usually much smaller than N. In the following, we also use the notations $\mathbf{x}_1$ and $\mathbf{x}_2$ in the sense of set. For example, when we say $i \in \mathbf{x}_1$, we mean that x_i is a state in the set $\mathbf{x}_1$.

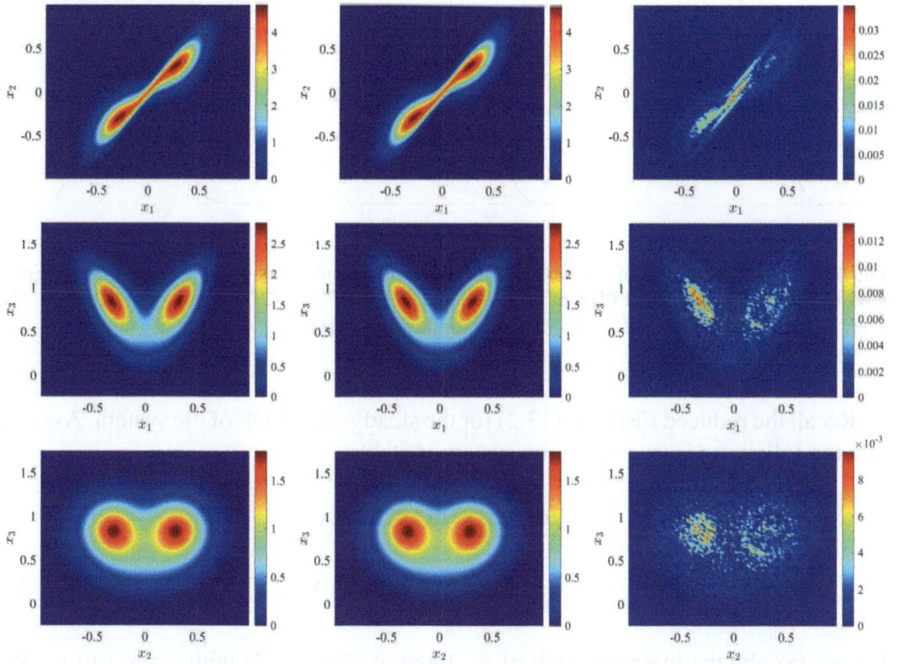

Fig. 13.2 The comparison of the RBFNN PDF solution and simulations of the 3D Lorenz system. Top row from left to right: (1) The RBFNN solution $p_{x_1x_2}$. (2) Monte Carlo simulation results $\bar{p}_{x_1x_2}$. (3) The absolute error between the RBFNN solution and Monte Carlo simulation results $\left|p_{x_1x_2} - \bar{p}_{x_1x_2}\right|$. Mid row from left to right: (1) The RBFNN solution $p_{x_1x_3}$. (2) Monte Carlo simulation results $\bar{p}_{x_1x_3}$. (3) Absolute error of $p_{x_1x_3}$. Bottom row from left to right: (1) The absolute error between the RBFNN solution and Monte Carlo simulation results $\left|p_{x_1x_3} - \bar{p}_{x_1x_3}\right|$. (2) Monte Carlo simulation results $\bar{p}_{x_2x_3}$. (3) The absolute error between the RBFNN solution and Monte Carlo simulation results $\left|p_{x_1x_3} - \bar{p}_{x_1x_3}\right|$

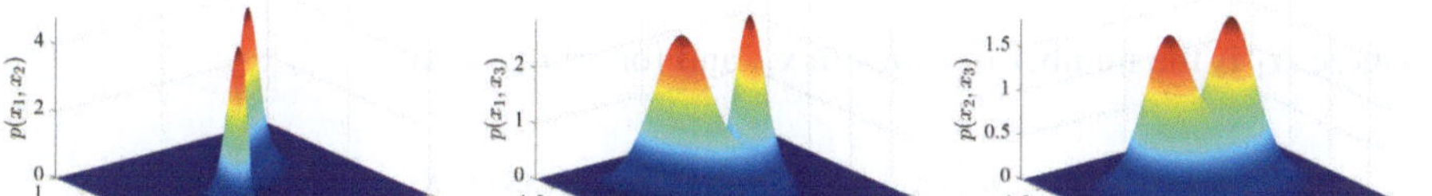

Fig. 13.3 The RBFNN solution of the 3D Lorenz system. From left to right: (1) The marginal PDF $p_{x_1x_2}$. (2) The marginal PDF $p_{x_1x_3}$. (3) The marginal PDF $p_{x_2x_3}$

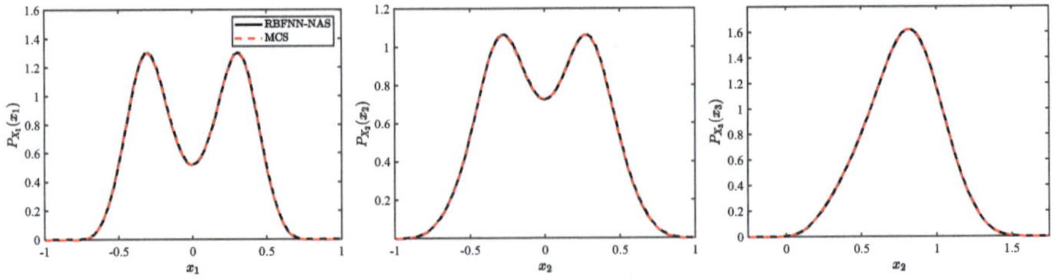

Fig. 13.4 The comparison of the RBFNN solution with simulations of the Lorenz system. From left to right: (1) The marginal PDF p_{x_1}. (2) The marginal PDF p_{x_2}. (3) The marginal PDF p_{x_3}. The agreement of the RBFNN solution with simulations is excellent

Recall the reduced FPK Eq. (13.2) for the steady-state PDF of the system. Assume that the following is true when the system is stable.

$$\lim_{x_i \to \infty} m_i(\mathbf{x}) p_{\mathbf{X}}(\mathbf{x}) = 0,$$
$$\lim_{x_i \to \infty} \frac{\partial}{\partial x_i} b_{ij}(\mathbf{x}) p_{\mathbf{X}}(\mathbf{x}) = 0, \forall i, j. \tag{13.31}$$

Let $p_{\mathbf{X}_1}(\mathbf{x}_1)$ be the marginal PDF of $\mathbf{x}_1$. Integrate Eq. (13.2) with respect to $\mathbf{x}_2$. We arrive the equation governing the marginal PDF $p_{\mathbf{X}_1}(\mathbf{x}_1)$.

$$-\sum_{i=1}^{N_1} \frac{\partial}{\partial x_i}[m_i^r(\mathbf{x}_1) p_{\mathbf{X}_1}(\mathbf{x}_1)] \tag{13.32}$$
$$+\sum_{i=1}^{N_1}\sum_{j=1}^{N_1} \frac{\partial^2}{\partial x_i \partial x_j}\left[\frac{b_{ij}^r(\mathbf{x})}{2} p_{\mathbf{X}_1}(\mathbf{x}_1)\right] = 0, \ i, j \in \mathbf{x}_1$$

where N_1 is the number of states in $\mathbf{x}_1$, and for all $i, j \in \mathbf{x}_1$,

$$m_i^r(\mathbf{x}_1) = \int_{\mathbb{R}_{\mathbf{x}_2}} m_i(\mathbf{x}) p_{\mathbf{X}|\mathbf{X}_1}(\mathbf{x}|\mathbf{x}_1) d\mathbf{x}_2$$
$$= \mathrm{E}[m_i(\mathbf{x})|\mathbf{X}_1 = \mathbf{x}_1], \tag{13.33}$$
$$b_{ij}^r(\mathbf{x}_1) = \int_{\mathbb{R}_{\mathbf{x}_2}} b_{ij}(\mathbf{x}) p_{\mathbf{X}|\mathbf{X}_1}(\mathbf{x}|\mathbf{x}_1) d\mathbf{x}_2$$
$$= \mathrm{E}[b_{ij}(\mathbf{x})|\mathbf{X}_1 = \mathbf{x}_1]. \tag{13.34}$$

$\mathbb{R}_{\mathbf{x}_2}$ denotes the space spanned by the states in $\mathbf{x}_2$.

13.2.2 Computing Conditional Expectation

Two conditional expectations for the drift and diffusion terms of the FPK equation for the conditional PDF of $\mathbf{x}_1$, i.e. $\mathrm{E}[m_i(\mathbf{x})|\mathbf{X}_1 = \mathbf{x}_1]$ and $\mathrm{E}[b_{ij}(\mathbf{x})|\mathbf{X}_1 = \mathbf{x}_1]$, usually cannot be obtained analytically. They are, however, relatively straightforward to compute with the help of Monte Carlo simulations.

Since the analytical expressions of the drift term $m_i^r(\mathbf{x}_1)$ and the diffusion term $b_{ij}^r(\mathbf{x}_1)$, i.e., the expectations, are not readily available as a function of $\mathbf{x}_1$, neural networks can be used to approximate them. RBFNN is used in [8] to approximate both the terms.

The training data from simulations include the estimates of $m_i^r(\mathbf{x}_1)$ and $b_{ij}^r(\mathbf{x}_1)$. $\mathbf{x}_1$ is also available because the expectation is conditioned on $\mathbf{x}_1$. Results reported in [8] suggest that a small number of samples can determine the neural network model of $m_i^r(\mathbf{x}_1)$ and $b_{ij}^r(\mathbf{x}_1)$ with sufficient accuracy.

Once the drift term $m_i^r(\mathbf{x}_1)$ and the diffusion term $b_{ij}^r(\mathbf{x}_1)$ are determined, another neural network solution is considered to estimate the conditional PDF $p_{\mathbf{X}|\mathbf{X}_1}(\mathbf{x}|\mathbf{x}_1)$. For more details and examples, refer to the Ref. [8].

References

1. Wang X, Jiang J, Hong L, Sun JQ (2022) Random vibration analysis with radial basis function neural networks. Int J Dyn Control 10(5):1385–1394. https://doi.org/10.1007/s40435-021-00893-2
2. Ye W, Chen L, Qian J, Sun JQ (2022) RBFNN for calculating the stationary response of SDOF nonlinear systems excited by Poisson white noise. Int J Struct Stab Dyn 2350019. https://doi.org/10.1142/S0219455423500190
3. Wang X, Jiang J, Hong L, Sun JQ (2022) Stochastic bifurcations and transient dynamics of probability responses with radial basis function neural networks. Int J Non-Linear Mech 147(104):244. https://doi.org/10.1016/j.ijnonlinmec.2022.104244
4. Qian J, Chen L, Sun JQ (2023) Random vibration analysis of vibro-impact systems: RBF neural network method. Int J Non-Linear Mech 148(104):261. https://doi.org/10.1016/j.ijnonlinmec.2022.104261
5. Qian J, Chen L, Sun JQ (2023) Transient response prediction of randomly excited vibro-impact systems via RBF neural networks. J Sound Vib 546(117):456. https://doi.org/10.1016/j.jsv.2022.117456
6. Wang X, Jiang J, Hong L, Sun JQ (2022) First-passage problem in random vibrations with radial basis function neural networks. J Vib Acoust 144(5):051,014. https://doi.org/10.1115/1.4054437
7. Wang X, Jiang J, Hong L, Chen L, Sun JQ (2023) On the optimal design of radial basis function neural networks for the analysis of nonlinear stochastic systems. Probab Eng Mech 73(103):470. https://doi.org/10.1016/j.probengmech.2023.103470
8. Xiao Y, Chen L, Duan Z, Sun JQ, Tang Y (2024) An efficient method for solving high-dimension stationary FPK equation of strongly nonlinear systems under additive and/or multiplicative white noise. Probabilistic Eng Mech 77:103,668. DOI j.probengmech.2024.103668

Index

J.-Q. Sun, *Random Vibration with Machine Learning Method*,
Nonlinear Physical Science, https://doi.org/10.1007/978-981-95-0812-9

MIX
Papier aus verantwortungsvollen Quellen
Paper from responsible sources
FSC® C105338
FSC
www.fsc.org

If you have any concerns about our products,
you can contact us on
ProductSafety@springernature.com

In case Publisher is established outside the EU,
the EU authorized representative is:
Springer Nature Customer Service Center GmbH
Europaplatz 3, 69115 Heidelberg, Germany

Printed by Libri Plureos GmbH
in Hamburg, Germany